Microarray Image Analysis

An Algorithmic Approach

Chapman & Hall/CRC
Computer Science and Data Analysis Series

The interface between the computer and statistical sciences is increasing, as each discipline seeks to harness the power and resources of the other. This series aims to foster the integration between the computer sciences and statistical, numerical, and probabilistic methods by publishing a broad range of reference works, textbooks, and handbooks.

SERIES EDITORS
David Blei, Princeton University
David Madigan, Rutgers University
Marina Meila, University of Washington
Fionn Murtagh, Royal Holloway, University of London

Proposals for the series should be sent directly to one of the series editors above, or submitted to:

Chapman & Hall/CRC
4th Floor, Albert House
1-4 Singer Street
London EC2A 4BQ
UK

Published Titles

Bayesian Artificial Intelligence
Kevin B. Korb and Ann E. Nicholson

Clustering for Data Mining:
   A Data Recovery Approach
Boris Mirkin

Computational Statistics Handbook with
   MATLAB®, Second Edition
Wendy L. Martinez and Angel R. Martinez

Correspondence Analysis and Data
   Coding with Java and R
Fionn Murtagh

Design and Modeling for Computer
   Experiments
Kai-Tai Fang, Runze Li, and Agus Sudjianto

Exploratory Data Analysis with MATLAB®
Wendy L. Martinez and Angel R. Martinez

Introduction to Data Technologies
Paul Murrell

Introduction to Machine Learning
   and Bioinformatics
*Sushmita Mitra, Sujay Datta,
   Theodore Perkins, and George Michailidis*

Microarray Image Analysis:
   An Algorithmic Approach
Karl Fraser, Zidong Wang, and Xiaohui Liu

Pattern Recognition Algorithms for
   Data Mining
Sankar K. Pal and Pabitra Mitra

R Graphics
Paul Murrell

R Programming for Bioinformatics
Robert Gentleman

Semisupervised Learning for
   Computational Linguistics
Steven Abney

Statistical Computing with R
Maria L. Rizzo

Computer Science and Data Analysis Series

Microarray Image Analysis

An Algorithmic Approach

Karl Fraser
Zidong Wang
Xiaohui Liu

CRC Press
Taylor & Francis Group
Boca Raton   London   New York

CRC Press is an imprint of the
Taylor & Francis Group, an **informa** business

A CHAPMAN & HALL BOOK

Chapman & Hall/CRC
Taylor & Francis Group
6000 Broken Sound Parkway NW, Suite 300
Boca Raton, FL 33487-2742

First issued in paperback 2017

© 2010 by Taylor and Francis Group, LLC
Chapman & Hall/CRC is an imprint of Taylor & Francis Group, an Informa business

No claim to original U.S. Government works

ISBN 13: 978-1-138-11515-6 (pbk)
ISBN 13: 978-1-4200-9153-3 (hbk)

Library of Congress Cataloging-in-Publication Data

Fraser, Karl.
    Microarray image analysis : an algorithmic approach / Karl Fraser, Zidong Wang, Xiaohui Liu.
        p. cm. -- (Computer science and data analysis series)
    Includes bibliographical references and index.
    ISBN 978-1-4200-9153-3 (hardcover : alk. paper)
    1. DNA microarrays. 2. Image processing--Digital techniques. I. Wang, Zidong. II. Liu, Xiaohui. III. Title.

QP624.5.D726F73 2010
572.8'636--dc22   2009044676

Visit the Taylor & Francis Web site at
http://www.taylorandfrancis.com

and the CRC Press Web site at
http://www.crcpress.com

Table of Contents

List of Figures

List of Algorithms

Preface and Acknowledgments

The paradigm shift heralded by the invention of DNA microarray technology has revolutionized the way in which biologists study the interaction and regulation of an organisms gene set. However, the technology is still in the early stages of design and development and there are, therefore, major challenges in the image processing phase of the analysis. Commercial microarray analysis systems typically rely on a human operator a great deal to complete the gene spot identification process. Any errors generated in the analysis stages will typically propagate through to the later stages of processing with the final analysis results not accounting for such errors. To harness the high-throughput potential of this technology it is crucial that the analysis stages of the process are decoupled from the requirements of operator assistance.

In this book, an automatic system for microarray image processing is proposed to make this decoupling a reality. The proposed system utilizes, extends and integrates traditional analytical-based methods and custom designed novel algorithms. In some cases, these traditional algorithms were incapable of directly processing the raw microarray imagery and therefore had to be optimized such that they would scale to these particularly large image datasets. In other cases, a novel method of clustering was proposed such that the underlying structure of a microarray image can be determined more readily via usage of the image's spatial and contextual knowledge. Such a contextual clustering process has potential benefits in other image analysis domains.

The book brings together the disparate fields of image processing, data analysis and molecular biology to make a systematic attempt in advancing the state of the art in this important area. It is envisaged that this automatic system will have significant benefits not only for microarray users directly, e.g., repeatable results and sounder analysis, but also for computer scientists and practitioners in other domains.

We take this opportunity to thank Dr. Daniel Morris who has made substantial contributions to Chapter 7. Special thanks to Dr. Paul Kellam, Dr. Suling Li, and Professor Joost Kok who between them provided data and other advice that has been invaluable. Thanks are also due to Dr. Paul O'Neill, Dr. Daniel Morris and fellow Intelligent Data Analysis Group members for their collaboration and support over the years.

Biographies

Karl Fraser received his BSc in computer systems & networks and his MSc in computational intelligence from Plymouth University, Plymouth, United Kingdom, in 2001, 2002 respectively. His PhD was awarded in computer science from Brunel University, London, United Kingdom, in 2006.

His PhD work identified a novel approach of analysing microarray images in order to improve quality and processing issues and not only showed strong evidence that current imaging techniques do not harness the available microarray information fully, but that such analysis can be done in a truly automated way. Dr. Fraser is currently a research fellow working on an EPSRC funded project entitled, "Reconstructing Background of DNA Microarray Imagery" within the Centre for Intelligent Data Analysis (CIDA).

Dr. Fraser's research interests lie in the fields of bioinformatics, image and signal processing, data mining, artificial intelligence, neuronal code generation, and dynamical systems.

Zidong Wang was born in Jiangsu, China, in 1966. He received a B.Sc. degree in mathematics in 1986 from Suzhou University, Suzhou, China, and a M.Sc. degree in applied mathematics in 1990 and a Ph.D. degree in electrical engineering in 1994, both from Nanjing University of Science and Technology, Nanjing, China.

He is currently professor of Dynamical Systems and Computing in the Department of Information Systems and Computing, Brunel University, United Kingdom. From 1990 to 2002, he held teaching and research appointments in universities in China, Germany, and the United Kingdom. Prof. Wang's research interests include dynamical systems, signal processing, bioinformatics, and control theory and applications. He has published more than 120 papers in refereed international journals. He is a holder of the Alexander von Humboldt Research Fellowship of Germany, the JSPS Research Fellowship of Japan, and William Mong Visiting Research Fellowship of Hong Kong.

Prof. Wang serves as an associate editor for 11 international journals, including *IEEE Transactions on Automatic Control, IEEE Transactions on Control Systems Technology, IEEE Transactions on Neural Networks, IEEE Transactions on Signal Processing*, and *IEEE Transactions on Systems, Man, and Cybernetics - Part C*. He is a senior member of the IEEE, a fellow of the Royal Statistical Society and a member of program committee for many international conferences.

Xiaohui Liu received the BEng degree in computing from Hohai University, Nanjing, China in 1982 and the PhD degree in computer science from Heriot-Watt University, Edinburgh, United Kingdom in 1988. He was appointed as professor of computing at Brunel University in 2000. Prior to this, he was a member of academic staff at Birkbeck College, University of London (1989-2000) and research staff at Durham University (1988-1989). Professor Liu is director of the Centre for Intelligent Data Analysis and has been research director for the School of Information Systems, Computing, and Mathematics at Brunel University since 2006.

Professor Liu is a charted engineer, life member of the Association for the Advancement of Artificial Intelligence, fellow of the Royal Statistical Society, and charted fellow of the British Computer Society. He has given numerous invited and keynote talks, chaired several international conferences, and advised funding agencies on interdisciplinary research programs.

1

Introduction

From its beginnings in a small Cambridge pub to the multi-billion dollar industry that exists today, discovering the secret of DNA has to be one of the driving forces behind modern scientific research. Over the years, research into this illusive code has taken many forms. Since the initial mapping and open publication of the human genome, one of the most publicized applications has to be that of microarray technology and its ability to facilitate the monitoring of many thousands of genes simultaneously.

Even though the landscape of data analysis has been seen to evolve rapidly in recent years, the basic desire to capture useful information within the data remains resilient. What was once a stronghold for the statistical community has been thrown to the field; tried and tested methods have failed to adapt to high dimensional, undersampled data whilst the largely applied domain of intelligent data analysis has begun to excel. No single community can now stake a claim to dominance, as it is becoming clearer that the only way forward is through a unity of techniques. Techniques developed for very different purposes, sometimes decades before, need to be rediscovered while novel ideas are developed. The challenge now is in utilizing this sometimes forgotten knowledge in environments and with data that is foreign to their original application, as well as, discovering appropriate techniques from fields that may have been previously disassociated. Only by focusing on their commonality, the interface that will allow the use of a technique designed originally for a completely different purpose, in a way never before envisioned, can the full potential of new vast sums of data be truly realized.

This book brings together the disparate fields of image processing, data analysis and molecular biology, along with many others to a lesser degree such as outlier detection, statistics and signal processing, all of which capture the essence of bioinformatics. Here the focus is on extracting improved results from existing cDNA microarray experiments, a technology that is still early in its development. The majority of researchers in this field rely on very simple techniques for the first stages of analysis and then use a range of statistical techniques to clean and filter the data. Instead, it is proposed that these early stages are the most crucial in terms of accuracy, as any errors found here will propagate down through all later stages of processing and analysis. With this in mind, work that has been conducted in the preprocessing of these early stages and a look at how this may be extended is presented.

1.1 Overview

From man's initial forays into all things, the one constant has been his need to quest. Be it pushing one's physical limits, or revolutionizing our understanding of nature. Man's thirst for knowledge is unquenchable. This inherent need to understand a "thing" fully, to push our theories and concepts into realms unknown, indeed, to make sense of the jumble of sensory and other phenomena we see in the world around us, pushes man to the edge of this very understanding. This nature of quest can be seen more obviously, perhaps, in the field of biological processes.

What differentiates man from the chimpanzees, why do humans all look similar yet different at the same time, how or why does a "normally" adjusted person become psychotic in nature? These types of questions "cry out" for answers, but unfortunately, our current knowledge is such that these answers take time to formulate and verify. The current push in this vast biological research domain is focused on genetics, be it the classical or molecular variety. Indeed, the distinction between the two subfields only started to appear about 56 years ago, with the work on the three-dimensional structure of deoxyribonucleic acid (DNA). Techniques used to study this DNA were still in the earliest stages of development and very little was understood (other than its structure). At this point classical genetics started to be defined as the study of how organism traits were transmitted in family histories. The problem was these families could not be compared directly, as genetic differences between the species were impossible to define (organisms of different species do not normally mate or their mixed offspring die early). The critical generalization was derived from the fruit experiments of Gregor Mendel, in which it is genes that are transmitted rather than traits. Classical genetics became interested in recognizing these alternative genes (through their effects on visible traits) and on mapping these genes along specific chromosomes. Molecular genetics, on the other hand, was more interested in questions such as how alternative genes differ and why these alternative genes express different traits. It should be noted, however, that there is no fundamental distinction between the two varieties; they are simply different approaches to understanding the genetic material of life.

Given the advances in modern genetics, it is possible to study the differences between species (and individuals) through the comparison and analysis of the relevant species DNA itself. With these improved DNA tools, biologists essentially examine the gene expression of an organism's interesting genes. The problem would be as to how, exactly, to identify these likely interesting genes from those available in the body. This caveat also assumes that it is known exactly where the individual gene sequences start and finish within the body's raw DNA sequence (which is currently estimated to contain 3 billion base pairs). The development of tools that can isolate and clone such large DNA fragments not only starts to yield insight into these answers, but also render the ability to map and sequence the human genome more accurately. Once such an accurate sequencing process is available, the practical benefits of these DNA techniques become more apparent. Knowledge of the genomes of pathogens, for example, may help with the development of vaccines for diseases. At a more abstract level, knowledge of the genomes of other model organisms (mouse and chicken for example) could reveal important new information about disease, neurobiology and other such biological processes.

As a direct consequence of these sequencing investigations, a new genetics approach called functional genomics has been born. This new field of genetics is interested in genome-wide patterns of gene expression and the mechanisms by which this gene expression is coordinated throughout an organism. As changes take place at the cellular level, the patterns of gene expression will also change. However, as genes are usually employed in sets (as opposed to individually), when one gene set's expression level is decreased, the expression level of a different coordinated gene set may well increase. How exactly then, should one go about selecting the appropriate gene sets for study? Should all genes in a given pathway be selected for the set, or alternatively, only those genes that render the pathways signature profile? This selection process is made all the more difficult by the problem of how the interactions of tens of thousands of genes can be simultaneously studied.

1.2 Current state of art

This simultaneous logistics problem was overcome with the relatively recent development of microarray technology [67, 142, 166, 205, 206]. Essentially, a microarray is a flat glass surface with the approximate dimensions of a standard postage stamp. The surface of the microarray is pre-treated such that background noise is minimized. The current technology [46, 61, 62, 104] makes possible the printing of between 10,000~100,000 gene spots in theory (into distinct grid-like groupings) onto the glass surface. Each spot represents a specific gene and contains the immobilized DNA sequence of said gene such that hybridization with the gene's appropriate DNA or ribonucleic acid (RNA) complement can take place. These hybridized sequences represent isolated cell lines grown under different conditions. It could be that one cell line is kept pure (no contamination), while the other cell line is exposed to a disease causing chemical, or represents different developmental stages of a disease such as cancer for instance. The hybridized sequences also include a marker or fluorescent label that facilitates the identification of the gene spot and their separation from the background of the slide at a later stage. Usually, a red label is used in the experimental sequence while a green label is used in the control sequence [186, 187]. The two samples are mixed and given time to hybridize with the pre-printed DNA present on the microarray surface itself.

As the samples are mixed to begin with, the hybridization is known as a competitive process, which means that the differences between the two experiment samples and the control or reference sample can be compared. The density of the resultant DNA strands bound to the microarray is said to be proportional to the concentration of the red or green molecules in the underlying gene spot sample [191]. Essentially, genes that are over expressed in the experiment sample relative to the reference sample will have more red molecules in their strands; while underexpressed genes will have more green molecules. These resultant gene spot expressions are indicative of the relative RNA present in the samples at a given time frame under the experiment conditions, which means that the expression levels yield insight into the underlying biological process at that time. Microarrays therefore are capable of assaying the relative expression levels of any messenger ribonucleic acid (mRNA) species as long as certain quality metrics are met. On completion of the hybridization stage, the microarray is washed and converted into its digital equivalent for gene spot analysis to begin.

The output of a microarray experiment process is a digital image [98, 197] containing tens of thousands of spots, where the spot's intensities represent the appropriate interactions of the underlying genes to a biological experiment condition. Therefore, if the gene spot intensities could be quantified, the biologist would gain an understanding of which genes are involved in a given biological condition (cancer propagation for example). Ideally, the only "objects" that will be present on the microarray surface are the gene spots themselves. Unfortunately, due to the biological processes involved and the nature of the engineered hardware, the microarray surface will typically be littered with multiple erroneous artifacts. Hardware-based artifacts can include grid displacement and uneven sample printing issues. Usually hardware-related artifacts are somewhat systematic or biased in nature and as such can be dealt with fairly easily. Biological artifacts on the other hand can include insufficient raw genetic material to begin with, poor protocol preparation issues or quite simply nature being nature. Due to their inherently unstable nature, biological artifacts are usually more difficult to rectify. How, for example, is it possible to differentiate between a true gene spot and a similar looking artifact area?

Traditional microarray based analysis systems usually employ a human operator to solve these similarity challenges by requiring the operator to manually (or semi-automatically) identify the valid gene spots in the image and demarcate them accordingly. With thousands of genes spotted onto a single microarray chip, it requires little imagination to realize that the demarcation of so many gene spots can take up valuable laboratory time for a researcher. More critically however, the researchers tend to introduce small gene spot edge accuracy bounding errors (via both the manual and semi-automatic processes). The effects of these operator created errors manifest themselves in the final analysis results as lower than expected gene spot intensities and poor experiment result repeatability. Due to the numbers of genes printed to the arrays, the identification of such operator induced errors can be difficult to catch before interpretation analysis has taken place. Also, whereas a non-microarray based process could simply re-run an experiment on detection of a faulty gene sequence for example, a re-run is impractical for the microarray based approach as the costs involved are much greater. Issues like experiment repeatability, costing considerations and potential throughput capabilities therefore suggest a truly automatic microarray analysis process would yield significant benefits to the biological community.

The term "high-dimensional" is open to subjective interpretation depending on the context of the data and the analysis techniques to be used. In the case of microarray images, essentially, there are twenty million–plus pixels with two observations per pixel. This represents a significant problem associated with automating a microarray analysis process. Automated techniques that determine representative gene spot expression values from this image data must, therefore, overcome various obstacles algorithmically if they are to be successful. Traditional approaches retain the human identification element such that determination of the gene spot pixels is appropriated from the outset. Algorithmic methods are then used to refine these selected pixels to locate probable gene spots throughout the full image surface. Well known adaptive shape techniques like seeded region growing [1] and watershedding [26, 208] have been used by traditional systems to help with these algorithmic processes. But good as they are, they still rely on the operator of such systems to provide the initial gene spot locations (at least approximately). In order to bring a fully automatic paradigm to fruition in this domain therefore requires novel processes that harness the advantages of unsupervised methods along with more traditional approaches.

1.3 Experimental approach

With the on-going trend in biology and many other fields of research into larger and larger datasets, traditional approaches to analysis are gradually losing their place on the podium. A look at statistics gives a prime example whereby high dimensional data with a low number of samples, often proves to be too challenging for many traditional techniques. Models can easily over fit the data and therefore no meaningful analysis can be conducted. Many statisticians blamed the experimental design, quiet rightly, as it produced data that could not be used. The experiments themselves, however, could not be changed as they were limited by many factors such as cost, limited experimental samples and design constraints. It was simply a matter of time before a change would transpire in analysis methods.

If someone says something cannot be done, there will always be an opposing view; someone who is metaphorically willing to jump off a cliff and then worry about how they might fly. In the case of data analysis, this role was eagerly taken by many computer scientists, ignoring those that said it theoretically could not be done, simply wanting to try it in a practical manor to see for themselves. It is in this climate, the field of Intelligent Data Analysis was envisioned, partly based on trial and error, as well as inventiveness, a statistical grounding and a reliance on a willingness to think beyond the scope of traditional data analysis. This work is not content to simply document previously devised theories, but instead looks to the development of smart algorithms to overcome new obstacles in data analysis, along with the application of existing techniques in inventive new ways.

An alternative for processing gene expression data is therefore explored. More specifically, focus is placed on mechanical cDNA spotted microarrays although this is not to say that technologies such as synthesized and protein-based arrays may not also benefit from a similar application of these techniques. From the creation of the initial arrays, current processing methodology can be viewed as a series of successive data reduction stages. Where each step reduces the data dimensionality, so that it may be further utilized but like many things in this world, it can be a double-edged sword. Could it be that the very techniques that are meant to be facilitating the analysis process may in fact be hindering it instead?

This book explores the current procedure for analyzing microarrays and proposes new techniques that can improve data quality. The approach used is always one of experimental curiosity. New ideas are tested, proving themselves on both synthetic and real data. Unlike the well-defined field of statistics, application of many of these techniques on trial, will not always give the expected results. Therefore vigorous testing is performed and some of the methods can and do fail. The time spent on even these failed techniques is still vital, as without these, the discoveries leading to notable improvements in the system may have been overlooked.

The very nature of the data mining presented in this work is that it will be applied to large datasets. It is not sufficient to prove an idea in theory on a cropped, possibly synthetic set of data. It is real world, usable results that count and this has been a consideration throughout all of the work in the subsequent chapters. Techniques that are prototyped on representative sections of the data are scaled up to processing full experiments. This involves both the distribution of processes across machines and the development of innovative solutions for scaling existing techniques.

Overall, this book can be seen to take an empirical approach, where research is conducted on actual data throughout the project and the results can be seen to be applicable to the problems in a practical sense. Work presented in this book is a chronological documentation of open-ended research conducted into microarrays over a period of years. Some experiments were important in that they guided the direction of later research, while others show the evolution of techniques from a simple proof of concept to their application on real data, such as the automated processing of many millions of data points, removing what was previously a manual process.

1.4   Key issues

The main aim of the book is to improve the processes involved in the analysis of microarray image data. Keep in mind, however, that these processes will be very similar to a broad range of medical and other computer vision analysis areas and as such the methods devised here could be applicable to other domains. Typically a microarray researcher will analyze the image surface to begin with to determine the positions of the gene spots as accurately as possible, with the internal algorithms (using these locations either directly or after further refinement) calculating various spot statistics. Essentially, these analysis processes are thus broken up into *data reduction, normalization, identification* and *statistical* stages, the purpose of which is to cope with the various image anomalies present. As mentioned above, a prominent feature with current microarray analysis systems is they usually require human intervention at the feature addressing stage. If this intervention stage can be minimized (and ultimately removed completely), and the subsequent analysis stages can be created accordingly, a way is opened for truly high-throughput microarray analysis systems.

This book focuses on the processes involved in the analysis stages of this digital image rather than the image's initial creation aspects. Therefore the study investigates the creation of a fully automatic analysis system. It is envisioned that the final process will represent a robust and accurate platform that is capable of processing a variety of microarray images irrespective of their underlying quality and structure. The study begins with the identification of the generic component parts required by a traditional system, GenePix [14,15] or Dapple [37] for example. These component parts set the requirements scope and define the objectives of the study. The remainder of this section presents a broad overview of the component parts as required in a generic microarray analysis process, be it fully automatic in nature or not.

Taken as a whole, the goal of the resultant four stages is to identify the gene spots within the microarray image and calculate an accurate representative value for the individual genes present. Once these representatives have been determined, the gene spots can then be compared between slides and different experiment runs to ascertain biological knowledge. If an error is introduced at any stage of the process, it will have an effect throughout the systems resultant analysis. There will of course be errors present in the data from the outset (background artifacts for example). Therefore, rather than creating processes that attempt to fix problems after the fact, attention should also be focused on fixing the problems as they occur. Such a "belt and braces" approach to the problem at hand should help keep errors grounded.

1.4.1   Noise reduction

Perhaps the most critical aspect of a microarray experiment process is the one of noise reduction. Due to the dynamics involved in an experiment run, it should be of no real surprise that noise elements are generated in the image. The task of these noise reduction processes is obvious, not so clear however is how this reduction can be achieved without introducing some negative effect onto the data that represents the features of interest. Dimensionality reduction techniques [24] usually sacrifice some form of data accuracy to render the problem smaller and therefore, computationally simpler. In turn, this smaller problem space can be analyzed more thoroughly, with generalizations usually made about the raw dataset. Although a typical microarray image will have a number of noisy data elements (non-gene spot pixels) far outweighing those of the genes themselves, the automatic identification of the gene spots is non-trivial. Packages such as GenePix overcome this noise identification problem by requiring an operator to manually mark the gene spot regions. Template structures are used to simplify the bulk of this identification process, with the operator focusing on the accuracy of the template members themselves primarily.

1.4.2   Gene spot identification

Once all of the noise (as is feasible) in an images surface has been removed, the analysis task becomes one of gene spot identification. A microarray image (from the test dataset) has physical dimensions of approximately 25×76mm, which in turn yields a digital image of 5000×2000 pixels. Of these twenty million–plus pixel observations (there are two image surfaces associated with an experiment) just over 2.2 million represent members of gene spots. This means the remaining pixels represent background noise and are of no analytical interest to the experiment. Whereas packages such as GenePix rely on the operator to identify these gene spot regions (and hence as a byproduct, identify background pixels), to remove the operator from the feature identification task completely would ensure that operator errors would not be introduced in the process. Such an operator removal process however raises non-trivial algorithmic challenges that need to be addressed for the processes to be successful.

1.4.3   Gene spot quantification

With a microarray's gene spots accurately identified, the analysis task becomes one of quantification. Here, the two or four microarray data channels for the particular gene spot in question must have their florescence intensity quantified in some way. This florescence is proportional to the amount of DNA that bond to the underlying gene region on the slide and therefore, is indicative of the level of activity of the gene in question. As previously highlighted, the pre-printed gene DNA on the slide represents a reference gene while that of the raw genetic sample material represents the disease or developmental stage under investigation. A method, therefore, is required to facilitate the comparison between the control and treatment genes such that a representative value is derived. These two cell lines are sampled at different wavelengths (during the digitization process) with the individual gene expressions usually captured as log_2 ratios. By dividing the expression levels of one channel by that of the other and applying a log calculation, it is possible to track an increase or decrease in the expression of a particular gene. The acquisition of these representative gene expression values in turn facilitates the ability to compare experiment genes to each other and other experiment runs. Such a log calculation is highlighted in

$$\text{Ration} = log_2 \left(\frac{\text{Control Intensity}}{\text{Disease Intensity}} \right) \qquad (1.1)$$

1.4.4 Slide and experiment normalization

Before any quantification process can take place, the underlying data elements should have some form of normalization applied to them. Ideally, normalization is the process whereby systematic bias or trend is removed from the dataset fully; however, this complete removal is rarely achievable in practice. Normally, amongst the gene spots of the microarray there are what are called "control spots". These "control spots" can be thought of as guide points that assist the biologist with correctly aligning the array during the scanning and gene spot identification stages of analysis. Another kind of "control spot" are called housekeeping genes, these usually consist of genetic material from an unrelated species, and serve as quality control tests. A good example of these control spots is the Amersham ScoreCard, which contains a set of 23 artificial genes. These genes generate pre-determined signals that do not change across samples or experiments. There are many varieties of normalization process including; global, spatial, local and dye swaps with various advantages and disadvantages for them all. As these "control spots" are susceptible to random noise (as per a normal gene), they must be spread throughout the array to be used effectively for normalization.

1.5   Contribution to knowledge

In current microarray analysis systems, much of the analysis work is focused on a manual or semi-automatic process. If these stages of manual intervention could be removed or at least minimized in their usage, without jeopardizing the quality of returned results, lab technicians will be more productive with their time. The creation of such automatic algorithmic based systems would also have significant benefits with regards to producing consistent results compared to the traditional approaches.

During the design of the proposed *Copasetic Microarray Analysis* (CMA) framework, new and existing algorithms were developed and improved accordingly such that the four key issues of the previous section could be addressed appropriately. Note that the word "copasetic" means "fine" and it is used in this context to mean that the analysis results are finer (more accurate) than other systems. Initially, a weakness was found to exist in the way in which existing microarray image analysis systems use the image data. Essentially, current systems fail to capitalize on the information provided by the two or more color channels of a microarray image. The novel research carried out in Chapter 3 proposed an *Image Transformation Engine* (ITE) that is able to clarify an image's feature space in an unsupervised manner. By giving more weighting to pixel values in the mid-to-high-end of the 2^{16} unique values, the ITE process is able to highlight pixels typically associated with gene spots (indirectly identifying the background pixels at the same time). If such an enhancement processes is integrated with a re-scaling from the 2^{16} possible values into the 2^8 range, not only are memory requirements and analysis time reduced, but if done correctly, knowledge loss is also minimized.

The algorithmic processes of the CMA framework are designed such that an overlap aspect between processes helps with the reduction of error propagation through the framework. Such overlapping functionality allows the overall goals of various processes to be retained with an added benefit of approaching problems from different angles. As already highlighted by the ITE process, taking multiple views of a problem can result in learning much more domain knowledge. Chapter 4 takes advantage of the overlap idea by exploring a novel clustering algorithm approach to the gene spot identification problem. The *Pyramidic Contextual Clustering* (PCC) technique facilitates the scaling-up of traditional clustering methods to larger datasets. PCC helps to improve on traditional clustering results by also utilizing an image's underlying contextual information (traditional clustering techniques discard this information). As opposed to the ITE processes pixel intensity interest, PCC focuses on the spatial distribution of these intensities. Importantly the PCC technique provides full transparency into the "internal decision process" of the underlying cluster method. Such transparency functionality allows clustering techniques to harness more dataset knowledge than would otherwise be possible.

Along with these so-called data preparation techniques, new methods have also been designed that facilitate the detection of a microarray image's grid-like composition in a blind manner. Chapter 5 explores techniques that determine the structural information from a combination of the ITE and PCC generated results. First, the *Image Layout* (IL) stage searches the image surface for inherently consistent structures. These structures are usually associated with the grid-like master block positions of the image's gene spots. When these master block locations have been appropriately determined, the *Image Structure* (IS) stage focuses in on the individual gene spot regions within the newly found master block areas and demarcate a gene spot's unique region.

Chapter 6 extends the result of the IS process by determining an accurate profile for the gene spots in the image via the *Spatial Binding* (SB) functionality. The SB process is a hybrid technique that utilizes the learned image knowledge from the previous stages along with new formulations to rebuild a gene spot's characteristic topology. With an accurate topology, the gene spot quantification process should thus yield results that are an improvement over traditional techniques (or at least similar in their value). Importantly, due to the CMA framework's modular design and inherent flexibility there is no reason why it should be limited in application to the microarray domain. 2D electrophoresis gel arrays have a similar structure (the grid-like structure is not so strictly defined) to microarray images and their analysis should not be so different. Affymetrix arrays are in principle very similar in structure to cDNA microarrays and the proposed framework may well prove to be an acceptable analysis platform in this domain.

In Chapter 7 it is proven that the developed approach for microarray subgrid detection is robust against high levels of noise, high percentages of missing spots and all of the other factors that complicate the task of microarray subgrid detection. The algorithm(s) tested are also proven to be potentially robust against future changes in microarray technology as they can cope with different shaped primitives, different subgrid configurations and subgrid distributions. The approach can also work with a range of image resolutions, offering time saving benefits and proving robustness against an increase/decrease in primitive size and scanner resolution. The proposed method is clearly more robust than the 1D projection based approaches and is more likely to be able to cope with any new development in microarray technology. The method may also offer application in other areas, as proven with the artificial image set, providing a method of identifying and locating regularly appearing groupings of the same high valued primitive within an image.

DNA microarray technology has provided an efficient way of measuring the expression levels of thousands of genes in a single experiment on a single "chip". It enables the monitoring of expression levels of thousands of genes simultaneously. After the microarray image process conducted in previous chapters, we would have obtained a global view on the expression levels of all genes when the cell undergoes specific conditions or processes. Obviously, the next step should be selecting a good model to fit gene regulatory networks is essential to a meaningful analysis of the expression data. In Chapter 10, we view the gene regulatory network as a dynamic stochastic model, which is composed of the gene measurement equation and the gene regulation equation. In order to reflect the reality, we consider the gene measurement from microarray as noisy, and assume that the gene regulation equation is a one-order autoregressive (AR) stochastic dynamic process. Note that it is very important to regard the models as stochastic, since the gene expression is of inherent stochasticity. Stochastic models can help conduct more realistic simulations of biological systems, and also set up a practical criterion for measuring the robustness of the mathematical models against stochastic noises. After specifying the model structure, we apply, for the first time, the EM algorithm for identifying both the model parameters and the actual value of gene expression levels. Note that EM algorithm is a learning algorithm that can handle sparse parameter identification and noisy data very well. It is also shown that the EM algorithm can cope with the microarray gene expression data with large number of variables but a small number of observations. Four real-world gene expression data sets are employed to demonstrate the effectiveness of our algorithm, and some indices are used to evaluate the models of inferred gene regulatory networks from the viewpoint of bioinformatics.

In Chapter 8, we examine the effects of applying both existing and new texture synthesis inspired reconstruction techniques to real-world microarray image data. It is shown that although the use of existing methods can be highly effective, their output quality and execution time with respect to microarray data needs to be significantly improved. To overcome such accuracy and timing challenges, we propose a novel approach to reconstructing a gene's underlying background by attempting to harness an image's global knowledge more intently along with the gene's neighbor pixels. The proposed technique takes advantage of the grouping concept of the frequency domain and characterizes the global entities in a systematic way. At the same time, a gene spot's local spatial knowledge is used to help restrict the spread of intensity within the constructed region. The results as obtained from the study show significant improvements over a commonly used package (GenePix). Specifically, not only was the gene repeat variance reduced from slides in the test set, but construction time was also decreased about 50% in comparison to O'Neill, and more than 75% to GenePix.

The issues related to extended edge problems of gene spot reconstruction are addressed in Chapter 9. An approach is proposed that utilizes a graph theory inspired pixel identification mechanism to select those pixels that are most similar to their direct neighbors within a pre-defined region. The highlighted pixel chains are then replaced according to their closest background region representative. The results show that the new method makes a significant improvement in gene expression reduction both directly and when compared with technical repeat variances. It is quite probable that a hybrid reconstruction system (able to classify to some extent a gene region) will be of great benefit to this analysis task. Such a hybrid system would use what is deemed to be the most appropriate reconstruction technique for a given gene. As we have now developed several separate reconstruction techniques, shown to be highly effective at their task, it is our belief that such a hybrid system can now be tackled appropriately as several reconstruction specific component parts are in place.

1.6 Structure of the book

Chapter 2 begins the study with an analysis of the current background material as is relevant to the book. This chronological breakdown of key discoveries made in the biological field not only presents the underlying techniques as they pertain to biology, but, lays the foundation for the introduction of the book's main subject matter - *Microarray Technology*. The paradigm shift in process presented by this technology is then examined and explained in detail. These details include but are not limited to; the creation of the microarrays themselves, the differences with the two main competing technologies, the generic microarray process and the required analysis stages of the resultant microarray output. The remaining chapters are focused on various aspects of this generic microarray process.

As will be highlighted, due to the biological processes involved in the generation of the microarray images, these resultant images do not have perfect separation between the features of interest (the gene spots) and their background. Indeed, a typical result for the digitization process will render an image which has lost features,* due to the variances involved. Chapter 3 presents a new technique that was designed to take advantage of a multiview approach to image analysis and thus yield clearer observations to these lost feature regions specifically, while more generally clarifying the full image area. The chapter starts by defining the precursor work to the final proposed solution. The solutions methods are then detailed with an emphasis on the underlying algorithmic workings. The results as rendered by these algorithms are analyzed over the application of real-world microarray image data. Finally the strengths, weaknesses and possible future enhancements of the proposed solution are highlighted.

As a typical microarray image contains in the order of twenty million–plus pixel observations, traditional clustering methods are unable to process the raw image directly. As such, research is needed in order to scale-up these and other methods to ever larger datasets. Chapter 4 addresses the challenges of applying some powerful traditional techniques - clustering in this case - to

---

*In most cases these features are not *lost*, but overshadowed by higher intensity features.

full-scale microarray experiments. The chapter starts by exploring the algorithmic processes and an analysis of these processes is presented over real-world microarray image data. The strengths and weaknesses of the technique are highlighted with future improvement considerations detailed.

Once these so-called data preparation stages have finished, the computation task becomes one of feature identification (the manual stage of packages such as GenePix). Chapter 5 details an approach to this feature identification task that has proved to be quite effective over the available microarray image sets. The chapter consists of two main processes which are both detailed independently. Essentially, a microarray is made up of multiple block groupings of gene spots spread over its surface area. The first analysis process is to determine the location of these so-called master blocks in the image. These master blocks in turn are made up of the individual gene spots. The second process must calculate a good approximation of these gene spot locations in order that later stages are able to separate the gene spot itself from its local background (at which point quantification can be made). After the chapter has presented a review of these general ideas, a detailed discussion of the two processes is made with appropriate evaluations over real-world microarray image data. The chapter rounds off with a discussion of the strengths and weaknesses of the two approaches and highlights possible future enhancements.

With the culmination of Chapter 5, the positions of the microarray images features are now known. Chapter 6 details a process that is able to examine these local image regions and quantify a very good approximation of the underlying gene spot. This process harnesses the knowledge learned from the aforementioned chapters and presents a technique that is able to render highly detailed surface models. As in previous chapters, the details of the algorithmic processes are presented along with an analysis of the processes performance over real-world microarray image data. The chapter ends with the presentation of the strengths and weaknesses of the technique with future enhancements mentioned as appropriate.

The spots printed onto a typical microarray slide are usually printed in separate subgrids each with the same number of rows and columns. The detection of these subgrids and the division of an image into its constituent subgrids is widely overlooked within the microarray image processing literature. As with spot detection, within Chapter 7 the task of subgrid detection will be discussed, with all issues that make this task difficult highlighted. Previous work in this area will then be discussed with weaknesses of past approaches identified. A new approach to this problem developed as part of this book is then described and tested. The results of testing this new approach show that it is an improvement upon all previous techniques and even offers applications within the broader topic of image analysis.

In Chapter 10, an expectation-maximization (EM) algorithm is developed for modeling the gene regulatory networks from gene expression time series data. The gene regulation matrix is obtained by an iterative learning procedure based on the stochastic linear dynamic model. Gene expression stochastic dynamic models for four real-world gene expression data sets are constructed to show how our algorithm works. Our algorithm can tackle the spare gene regulatory networks only by setting some of the matrix entries as zero. Furthermore, our algorithm is especially efficient for lager gene regulatory networks which can be divided into several individual gene regulatory networks. Therefore, our algorithm can be ideally applied to modeling the gene regulatory networks where the real connectivity of the network is specified a priori. We expect that our algorithm will be useful for reconstructing gene regulatory networks on a genome-wide scale, and hope that our results will benefit for biologists and biological computation scientists. In the near future we will continue to investigate some real-world gene expression data sets and apply our algorithm to reconstruct gene regulatory networks with missing data, sparse connectivity, periodicity and time-delays. We are also getting connection with the biologists to explain our results from the biological point of view.

Although microarrays allow biologists to determine gene expressions for tens of thousands of genes simultaneously, their biological grounding results in output images that are permeated with noise. During gene expression quantification, a gene's "noise" or "background" must therefore be removed for purposes of precision. Chapter 8 presents a novel technique for such a background removal process. With use of the newly generated background image, the gene expressions can be calculated more precisely. The technique's ability to account for the influence of outliers and other such artifacts more appropriately also means that the final gene expressions have considerably reduced variability. Experiments are carried out to test the technique against a mainstream and an alternative microarray analysis method. Our process is shown to reduce variability in the final expression results.

Microarrays produce high-resolution image data that are, unfortunately, permeated with a great deal of "noise" that must be removed for precision purposes. This chapter presents a technique for such a removal process. On completion of this non-trivial task, in Chapter 9, a new surface (devoid of gene spots) is subtracted from the original to render more precise gene expressions. The Graph-Cutting technique as implemented has the benefits that only the most appropriate pixels are replaced and these replacements are replicates rather than estimates. This means the influence of outliers and other artifacts are handled more appropriately (than in previous methods) as well as the variability of the final gene expressions being considerably reduced. Experiments are carried out to test the technique against commercial and previously researched reconstruction methods. Final results show that the graph-cutting inspired identification mechanism has a positive significant impact on reconstruction accuracy.

Chapter 11 discusses the contributions of the work presented in this book and describes the overall benefits of the proposed system in the biological and computer science domains. Future considerations of new research and related ideas are also discussed.

Finally, Chapter 12 is about the appendices on the background knowledge of microarray variants, basic transformations, clustering, mining gene expression data, autocorrelation and generalized "circular" Hough transform.

2

Background

2.1 Introduction

Modern day genetics is typified by sequencing projects [13, 77, 83, 102, 138, 144, 223]. Although not as laudable as that of the human genome project itself [108], the sequence projects still represent a significant jump in the human quest for knowledge. These projects also represent the cornerstones of modern biologically orientated knowledge through which the complexities of life take shape. The continued advancements in the illumination of these biological processes will have profound consequences on the future of humanity, improved health care and greater quality of life through better drug creation for example.

However, the underlying knowledge required for these significant biological advancements did not suddenly coalesce into the general consciousness. It has taken many years for eminent scientists of their day to slowly tease the fundamental genetics concepts and details out from the darkness. In hindsight, concepts like trait inheritance which are passed on from parent to child could appear to be quite obvious, but at the time could well have caused a similar stir with the establishment as that of Galileo's* work. Inheritance is a fundamental concept of life and integral to the functioning of evolutionary processes. As the mechanisms of inheritance are understood in greater detail, the techniques available to exploit inheritance behavior will also improve.

Microarrays represent the latest research tools that try to exploit the behavioral characteristics of the inheritance mechanisms on a large scale. Generally, microarrays are small glass, silicon or nylon membrane slides that give biologists the capability to simultaneously analyze many thousands of genes. Due to the massive shift in process capability, microarray systems will have a dramatic effect on the future of pharmacological based studies.

In order to give the reader a clear understanding of microarray technology and the biology that is relied upon for the process to be successful, this chap-

---

*Galileo (1564-1642) was the first astronomer to make full use of the telescope and observed the craters of the moon and the satellites of Jupiter. His open advocacy of Copernican cosmology however, led to a clash with the Catholic Counter-Reformation, and he spent his final years under house arrest.

ter focuses on providing the background for the study. In Section 2.2, a brief review of genetics history is offered such that the underlying concepts and techniques microarrays rely upon for their functionality are understood. Section 2.3 provides a brief summary of microarray technology itself which should give the reader an appreciation of the challenges involved in this interesting computer vision domain. The section introduces a typical microarray usage scenario with the key phases of the microarray process highlighted. After the general discussion of microarray processes, the review then details the three main phases involved in the analysis stage in Section 2.4. Section 2.5 briefly discusses a framework that has been proposed as a holistic solution to the issues as discussed in Sections 2.3-2.4. A summary of the chapter is presented in Section 2.6.

2.2   Molecular biology

As is now known, the inheritance carrier is in fact deoxyribonucleic acid (DNA). Similar to the discovery of the inheritance material itself this DNA structure required rigorous research over many years. The key events of these investigations are highlighted and discussed with the basic details of the code of life described.

2.2.1   Inheritance and the structure of DNA

Although molecular genetics deals with the study of a gene's structure and function at the molecular level, classical genetics (the precursor) was not aware of the existence of genes and their influence on life. However, classical genetics scientists did theorize that all living organisms must contain something - some genetic information - that was responsible for the familiar characteristics that arise between a parent and their child. This perspective was proposed long before the concept of "inheritance" was introduced, and the eventual realization that DNA and ribonucleic acid (RNA) carried the so-called code of life in their chromosomes.

In 1952, A. D. Hershey and M. Chase [101] extended research in virus reproduction mechanisms and conducted a series of experiments that would prove beyond doubt[†] whether it was protein or DNA that contained the hereditary material of life. Hershey and Chase concluded that DNA was in fact the physical carrier of heredity information.

[†]Awarded the Nobel in 1969 along with M. Delbruck and S. Luria for their work on the replication mechanism and the genetics of viruses.

At around the same time as the Hershey and Chase studies where underway research teams were also trying to determine the structure of DNA. DNA and RNA molecules where established as polymers[‡], the monomers of which are in turn called nucleotides. Each nucleotide consists of three distinct parts: a pentose sugar, a nitrogenous base, and a phosphate salt group. The sugar differs in DNA and RNA bases as deoxyribose and ribose respectively. The nitrogenous base consists of two classes: purines (consisting of adenine (A) and guanine (G)) and pyrimidines (consisting of thymine (T), cytosine (C), and uracil (U)). In RNA, uracil (U) occurs in place of the thymine. The phosphate group normally contains anion PO_4^3 or the $-OPO(OH)_2$ group.

However, the composition of DNA did not directly elucidate the role as carrier for inheritance and ultimately the gene. The true importance of the base pairs was to pose the problems; how exactly is the DNA sequence information useful and how can this simple four code sequence direct the replication process required by life. For replication to be viable the DNA information must be chemically stable and relatively static. However, for the processes of evolution to function correctly DNA information must also be capable of mutational change.

Many scientists focused on deciphering the structure of DNA which at the time was perceived to have stringent requirements, amongst the scientists were J. Watson, F. Crick, R. Franklin, and M. Wilkens who gathered all of the available DNA data. This data included the works of Pauling [160–162] who established proteins where helically coiled, Chargaff [40, 41] who determined base-pair balancing and the x-ray crystallography of Franklin and Gosling [81, 82]. This work contributed greatly to the research team of Watson and Crick [215, 216] who eventually determined the model of DNA to be a double helix configuration, as shown on the right of Figure 2.1, with bases at the center of the helix and sugar-phosphate units along the sides. The strands are complementary in nature (as shown on the left of the figure) that means adenine only pairs with thymine and guanine with cytosine.

This in-turn means, therefore, that if the base sequence of one strand is known, so is the base sequence for the other strand (the complement). There is no restriction to the sequence of bases in a single strand, so any sequence could be present along a strand. This model answered the initial concerns with respect to the given inheritance issues as follows:

- DNA could contain the genetic information in a coded form in the sequence of bases and therefore a cell's structure and function could be built from this sequence.

- The sequence of bases could be copied by using each "partner" strand as the pattern for a new partner with a complementary set of bases.

[‡]A substance with the molecular structure formed from many identical small molecules bonded together.

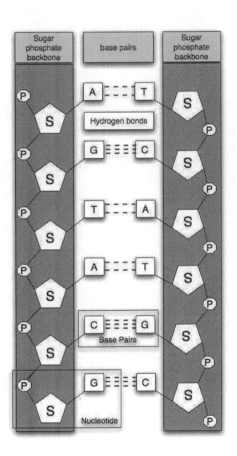

FIGURE 2.1
The structure of DNA.

- Changes in genetic information could result from errors during copying the base sequence.

With the DNA structure in place and a basic understanding of the composition starting to take shape, scientists also started to focus on how, exactly, a double helical structure coded is for genetic replication and function [216]. Essentially, the genetic code came to be characterized as the relation between a set of three bases — a codon — and one of twenty amino acids, the building blocks of proteins. This fundamental principle became known as biology's central dogma.

2.2.2 Central dogma

The central dogma is the fundamental principle of molecular genetics because it describes how the genetic information in a DNA sequence is expressed. In a sequence of DNA, the genetic code for protein synthesis contained in the strand is decoded in a linear fashion. A typical protein is made up of one or more polypeptide[§] chains with the chain members being called codons. There are twenty different amino acids coded for by the four base pairs. Each of these codons consists of three adjacent pairs, for example "UUU" specifies phenylalanine,[¶] which yields sixty four possible amino acids. However the extra combinations code for duplicate amino acids which give the code an elegant level of robustness. These codon groups then become genes when read from the correct starting locations with the genes in turn creating specific proteins.

DNA codes for protein indirectly however, via the processes of transcription and translation. These processes are known as the central dogma [49–51, 214] of molecular genetics. The central dogma cycle repeats itself on a system-wide scale throughout every living organism. If a sample of RNA could be captured, it would indicate which genes had been in use at that point in time, such information would yield great insight into the processes of life. Southern Blots bring the ideas of DNA cleavage, electrophoresis, nitrocellulose transfer and hybridization together, and through these various methods, scientists were able to take snapshots of the internal gene communication processes. The technique was developed in 1975 by E. M. Southern [195] and is, in essence, capable of locating a sequence of DNA within a complex mixture. Similar techniques have since been developed to locate a sequence of RNA (Northern Blotting) and for locating a protein (Western Blotting). It is important to note that these techniques find singular genes, sequences or proteins rather than sets of them.

[§]Organic polymer consisting of a large number of amino acid residues that form all or part of a protein.
[¶]Discovered by M. W. Nirenberg and H. J. Matthaei [154].

However, the techniques mentioned above are not powerful enough to permit a full understanding of multiple sequences within one step (as living organisms tend to have many thousands of genes), which would be more advantageous for biologists. The significance of such an understanding can be seen with the appearance of a technology called microarrays that would revolutionize the biotech industry.

2.3   Microarray technology

It is obviously valuable to have a general understanding of some of the underlying genetics. Most people will be familiar with deoxyribonucleic acid, also known as DNA. The code store in its structure forms the blue print for all living organisms on this planet. Simple pictorial representations of this often take a stylized view of its real structure, presenting a twisted ladder like form to represent the double helix molecular backbone. In this view, the basic code is stored by variations of chemicals in the "run's" of the ladder. The code itself is not simply a single long sequence of instructions but can be subdivided into smaller groups. The larger sections of sequences are referred to as chromosomes, these in turn, can be divided into smaller portions called genes. A gene itself is a length of code that specifies the blueprint for building many of the essential components for life such as an amino acid and a protein.

2.3.1   Gene expression

In an attempt to understand this code, biologists have moved into the study of gene expression. It provides an insight into the inner workings of a cell, by exploiting the fact that when a cell needs to react to a given situation it will normally do so by manufacturing the materials it requires. In order to do this a cell uses the correct sequences of code for each task and this is stored in the genes, which are in turn housed in the nucleus of the cell. A parallel can be drawn with the use of a library, where if someone requires information about a specific topic they would then have to find the correct book from the library before utilizing the information within. The cell's life, however, depends on the information stored in the nucleus and so unlike a library the information cannot be freely removed. Therefore, the analogy of a library can be adapted by a simple change: All the information is stored in the same way; but in order to make use of a particular book, it would have to be internally photocopied and then only this copy would be allowed to leave the premises. In this way, all of the material in the library is protected in the same way as DNA in a cell nucleus. In a cell, the information stored in the DNA is copied in the nucleus using another molecular structure called RNA and it is this, which is permitted to leave the nucleus and used from this point onwards.

This distinction is important when considering methods to analyze what is taking place in a cell. In the previous example of the library, if a method was devised to capture all of the photocopied material at a given instance, then some conclusions about the use of the library could be drawn. By repeating this over time or after key events, it may be possible to see how the use of the system is changing. In the same way, collecting all of the RNA that is present in a cell at a given point in time will allow the sections of DNA that are in use to be identified. This is important as changes to the cells environment, such as the introduction of cancerous material, may prompt different responses. Understanding these underlying mechanisms in this way would lead to new avenues of drug development. Extracting the RNA from a cell, however, is only half the problem, as crucially the sections of DNA that are in use in this RNA must be identified.

This introduces another problem, as the biology of a cell or for that matter, any living organism is not an exact science. It is impossible to define the use of a single gene in terms of true and false. There can always be the chance of residue RNA in a cell or similar sequences of RNA, which will prove to provide false positives. To complicate issues, a simple quantification of the amount of RNA representing a particular gene is also misrepresentative of its use, as this can vary over time. So without the ability to set a threshold level to specify the presence of a gene and with no meaningful way to quantify the level of a gene, how can the changing dynamics of a cell be measured?

The answer to this quandary can be found in the use of a competitive hybridization experiment. In its simplest form, this compares the state of a gene in a treated cell sample to that of a reference. This will then allow a gene's expression to be quantified in terms of the change between the two. For example, a gene in the treated cell line may be twice as abundant as that in the control cell line; in which case this may indicate a change to the internal workings of the cell. This can be important when genes form part of a pathway, a pattern of expression, which if blocked at a certain point by isolating the correct gene can lead to alterations in the cell's function. A virus is a good example of this, whereby, if the mechanism for infection can be stopped by inhibiting the expression of a particular gene then this may lead to the development of new treatments.

The human genome contains many thousands of genes and so looking at a single gene in isolation is not an option. Therefore, a method is required to facilitate comparisons between genes, combining both the control and treated cell lines into one representative value. The most common way to achieve this is with a log_2 ratio. By dividing the expression measurement in one channel by that of the other and then applying a log transform, it allows both an increase and decrease in the expression of a gene, to be compared with other genes from both the same and alternative experiments. This allows for the possibility of many genes to be compared to one another in unison and all that is required is a technique in which many more genes can be analyzed.

2.3.2   Microarrays

Prior to microarray technology, techniques like southern blots were used exclusively to analyze the gene expression of a few isolated genes. The southern blots analysis offered an overall picture of an organism's activities slowly taking shape. Southern blot techniques worked well, but they where wholly inadequate for the task of understanding life, especially given the recent successes of activities like the human genome project. The need for developing more advanced analysis tools lead to a fundamental shift in the biology literature, which focused on presenting the results garnered from thousands of genes rather than several.

The emergence of microarrays is typically traced back[||] to the work of Fodor [78, 79, 163] and later Brown's [186, 187] team at Stanford University, who** worked with others as consultants to Affymetrix (see Appendix A for details of current array systems), developed the basic technology for producing complementary DNA (cDNA) microarrays. By definition, a microarray is an ordered arrangement of samples or spots. Microarrays are usually described as the DNA slide or chip which consists of gene spots of approximately 300 microns in diameter or larger. DNA microarrays however generally apply to DNA chips whose gene spots have diameters of 200 microns or less. Within the microarray community there is some divergence between the terminology associated with the pre-printed DNA and that of the labeled or cDNA. Throughout this book, the pre-printed DNA will be referred to as probes, while the cDNA will be known as targets.

With the mass production of microarray data, a new problem has evolved, whereas before a single experiment could be analyzed manually, easily identifying and possibly correcting for lab-based errors. Now, this is just infeasible on the large number of gene experiments in production. Not only this, but if an error is identified in a gene, it is not just a case of re-running the experiment, as the costs are just too great. This is compounded by the complexities of analysis, as the manual techniques used before have given way to automated processing which is unable to treat each of the many thousands of genes on an individual basis. This leads to an automated system where simple methods are used to reduce the data into a usable form before the application of various filtering, normalization and other statistical techniques in order to correct for anomalies in the data. The reality of this is a large percentage of the data is discarded due to processing errors and simply because the basic techniques used in the early stages of processing are unable to deal with relatively simple defects in the data. When everything is accounted for, even with the current

---

[||]Which technically started to appear in the ligand-assay field, during the mid to late 1980s. [71, 72, 74]

**The Stanford Lab work relied on that of previous scientists. Fred Sanger and Walter Gilbert [179–182] pioneered modern DNA sequencing methods. Kary Mullis [145] discovered the polymerase chain reaction (PCR) method. Leroy Hood [189] described a method of DNA sequencing using fluorescent tags.

problems in analysis, microarrays are still an invaluable tool for gene expression analysis and work on improving their use will benefit both current and future research.

Microarray technology is a revolution in terms of the insight it provides into previously unexplored biological systems. With this ability to peek into the underlying life processes of all living organisms, it has a potential that cannot begin to be quantified, only hindered by its current immaturity. A relatively short span of a little over fifty years has seen the field evolve from the first publication of the structure of DNA to the draft mapping of the human "blue print" at the turn of the millennia. No one can ignore the potential impact this will have as a new era in our discovery of the very building blocks of life has begun, although, undoubtedly there are going to be a few teething problems.

A clear analogy can be drawn between these new techniques and the early development of computer systems and programming techniques throughout the 1960s and 1970s. During this period, many of the general programming practices that are now adopted to keep code clear, understandable and maintainable just did not yet exist. This led to an un-marshaled style of development whereby many problems were overlooked and many others were solved with solutions which should never have been anything more than a temporary fix. This kind of development worked and evolved for many years, but some of the problems just remained unresolved, waiting for a chance to resurface. One example that received a lot of media attention is that of the "Y2k" or millennium bug as it has become known. Essentially, a small oversight in the storage of dates that resulted in many billions of dollars spent worldwide on correcting the problem. Although this is an extreme example, it does highlight the need to be vigilant at all stages of a process, simply trouble shooting is not enough.

In much the same way microarrays are developing so quickly that often techniques and protocols are implemented and adopted in a fire-fighting manor. When there is a problem, it is temporarily patched up in order for the technology to progress and although the fix may seem to work, the underlying cause may remain a mystery. A possible example of this can be seen in many of the normalization and background correction techniques such as lowess [227] and averaging replicates [168], which commonly appear in the literature. Although they seem to correct for certain problems with the data, no extensive research has been conducted into the possible effects the technique may have and to see if the technique is really correcting the problem or simply obscuring it. An extreme example would be a technique that could be proposed that would normalize two datasets so that a series of repeats would show improved correlation. Such a technique need be nothing more than an algorithm that outputs a repeated sequence of values with a small variance; of course, this has invalidated the data and would be of no practical value. The point is, without understanding the consequences of a technique, it is very easy to get a processed result that appears correct.

The biggest problem with microarray data is that to get a set of results that can be analyzed, the original raw data has to traverse a veritable minefield of processing techniques, any one of which could be introducing error. Typically, the initial analysis of a microarray includes stages of digitization, gene spot identification, local background noise correction, gene spot summation, slide based filtering and normalization, and experiment based filtering and normalization.

The overall goal of these stages is to reduce what starts out as a series of pixels that fall within the bounds of a gene spot, into a representative value for that gene which can then be compared between slides and possibly between different experiments. As with any pipeline of this nature, if an error is introduced at an early stage, it will propagate through to the end of the process. It may possibly be amplified even further by later stages. This leads to at best a gene which would have been a valid result being filtered from the final set of data and a worst case of the error adversely affecting the analysis which is performed on the data.

One conclusion is that each stage of analyzing microarray data needs to be developed with a mind to process the data as accurately as possible. Particular care needs to be adopted throughout the early stages of pre-processing, as any errors introduced here will propagate through to the final output. Finally, there will always be errors that are already present in the data, such as random background noise; the initial stages of pre-processing should take every care to limit the effect of this at an early stage. This allows focus to be placed on the areas of real interest, the gene spots themselves. Rather than attempting to filter out problems long after they have been introduced, instead can a look back to the earlier analysis minimize the root cause?

In order to further understand the microarray process, it is important to identify the key phases of a typical microarray usage scenario.

2.3.3 Process summary

A common scenario for a microarray experiment can be broadly split into two key phases as follows:

- *Hybridization Preparation* - The slides are washed with a specially formulated substance that allows the DNA to bind to the glass surface. The probes of known reference genes are then printed to the slide in a pre-determined scheme. The cells of interest are given time to grow and settle into their environment (the injection of cancerous material into a laboratory rat, for example). Once this genetic material has settled into its environment, a sample of cell material is taken from the rats in question (control and treatment sets). These samples are spun independently by a centrifugal device in order to remove any liquid medium. The remaining messenger ribonucleic acid (mRNA) is then extracted. However, this extracted material needs to be greatly increased for it to be usable in a practical sense; this replication is accomplished by a technique called polymerase chain reaction (PCR) [145].

Once there is sufficient genetic material for the microarray process, the nucleotides of the mRNA sequence are fluorescently tagged for tracking purposes and the sequence is renamed cDNA (target) (see Appendix A for details of different chip manufacturing processes). In so-called two-dye microarrays these tagging agents are known as Cyanine 5 (Cy5) and Cyanine 3 (Cy3). Their frequencies are such that they are normally shown as red and green hues in resultant imagery respectively (as seen in the representative output image in Figure 2.2). Both sets of samples can then be thoroughly mixed in preparation for the hybridization process. Nucleic acid hybridization is the step in which the target and probe DNA material on the slide form heteroduplexes via Watson-Crick pairing. Generally, during hybridization the higher the salt content and the lower the temperature, the greater proportion of mismatched nucleotide base pairs will be generated within a given pairing sequence.

- *Digital Generation* - Once the hybridization process has been given time to complete, the slides are washed to remove any unbound target from the genes (this washing also helps remove excess and unbound material). The completed slide is then converted (the scanner in Figure 2.2) to gather the gene spot information into a digital format. The scanner device traverses the full slide surface with a green and red laser separately in the case of the two-color scenario recording the reflected light from the target/probe molecules. The most common scanning techniques employed either have a charge-coupled device (CCD) or a laser generator at their core (see Appendix A for different scanning processes).

The aim of the two phases above is to permit the comparison of activation or non-activation of a population of genes. On the basis of this digital information a computer can analyze (Section 2.4 discusses the analysis process in more detail) the interaction relationships between the control and treatment members, allowing the biologists to look at these snapshots of biological system activity and deduce internal interaction processes. In the case of two-color microarray processes (see the representative output image in Figure 2.2), gene spots of a solid green hue represent pure control and spots of a solid red hue represent pure treatment bindings respectively. Gene spots with a yellow hue represent a balanced binding between the two sets (control and treatment) over the specific gene.

2.3.4 Final output

As mentioned above, a variety of different microarray processes exist, each with unique benefits and challenges to be overcome. The remainder of this section provides a brief review of the differences the hardware approach makes

FIGURE 2.2

Common cDNA microarray scenario process and a representative output image

to the final microarray output image. The output images as shown in Figures 2.3-2.4 contain the same kind of genetic information, but due to compositional differences, their individual analyses will present different challenges. The noise elements of these different image surfaces are also outlined. Note the term "noise" is used in the generic sense to mean unwanted signal. Noise, artifact, contamination, etc., are representative of a similar phenomenon in this context.

As can be seen from Figure 2.3, a typical glass slide chip consists of thousands of gene spots distributed into distinct blocks across the surface. This image measures approximately 5000×2000 pixels (with 9216 gene spots) in a physical area of 25×76mm. The image requires 40 megabytes of storage memory. Due to the construction methods used and the inherent variation which can be expected as part of any biological process, these images are extremely noisy. This can especially be seen in the zoomed section that shows a selection of the type of problems that are typically encountered: missing or partial gene spots, morphological inconsistencies, alignment issues and background variation. Although there are several inconsistencies, two obvious artifacts can clearly be seen in the upper left and lower right quadrants of the un-zoomed image. Both artifacts in this case represent washing defects, but the one in the lower right quadrant is obviously obscuring several gene spot positions.

Figure 2.4 shows a typical experiment as printed onto a nylon membrane surface. Here the surface measures 1050×1050 pixels (containing ∼6000 gene spots) with a physical size of 40×30mm. As above, there is an element of biological variation within the image, but the bigger problem would be that of bleeding. Some of the larger features quite clearly merge with their immediate neighbors.

The photolithographic (Affymetrix) surface of Figure 2.5 measures 3650×3650

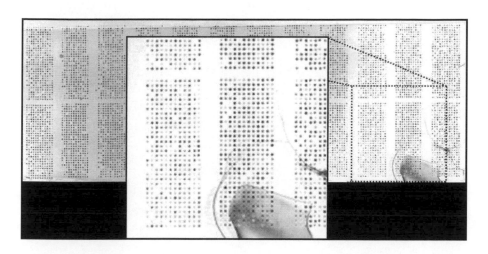

FIGURE 2.3
Example microarray process image output: glass slide cDNA

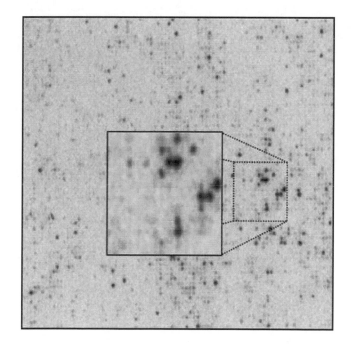

FIGURE 2.4
Example microarray process image output: nylon membrane cDNA

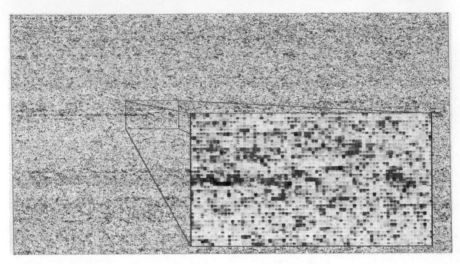

FIGURE 2.5

Example microarray process image output: photolithograph oligonucliotide

pixels and 12.8×12.8mm in physical terms. The gene spots in this image are rectangular in shape and are aligned as match and mismatch pairings. As photolithographic techniques typically use a mask structure to block light, feature leakage occurs within the match/mismatch pairings. Analysis software used on these images typically quantifies the gene expression via the internal region of the features rather than their whole area.

Irrespective of the microarray hardware used however, all of the images contain gene spots or features that need to be quantified in some way. Note also that the quantification process is made more difficult by the noise (or contamination) that is inherently present in the microarray creation process. Once such digitized images are available, the analysis process itself can begin.

2.4 Microarray analysis

This section focuses exclusively on *glass slide* cDNA microarray image analysis. The terms and concepts described here will be applicable to other microarray methods (as mentioned above and outlined in Appendix A) in general. Once the microarray image has been converted into the digital domain, the process of analysis can begin. As discussed previously, the microarray device was designed to allow biologists to perform thousands of simultaneous experiments at a time.

When these experiments have been performed they need to be quantified in such a way as to identify the gene expression of the individual pre-printed genes and thus acquire the transcription abundance for said gene. With this transcription quantity in hand, biologists can begin to infer the activities of the associated genes as they will be up or down regulated with respect to each other, with the up/down regulation relationship equally respected. These transcription values from the image analysis stage are commonly focused at finding similarities between the behavior of known and unknown genes in the initial samples. Such behavior analysis renders the ability to predict the function of unknown genes and helps to filter predictable gene functions such that any remaining or interesting gene characteristic can be analyzed further.

The image analysis processes overall goal is to find the genes within the imagery to be quantified; such a search process can thus be broken up into three main categories of *addressing*, *segmentation*, and *feature extraction*:

1) Addressing - a typical slide has several gene groupings on its surface, with hundreds of genes per group. Addressing, as its name implies, uniquely identifies the locations of the genes in the image in such a way that there is no overlap between any two gene locations.

2) Segmentation - once the genes have been localized by the addressing process, the task becomes one of accurately separating the gene spots themselves from their local background area.

3) Feature Extraction - feature extraction quantifies the area of interest's intensity measure along with the features local background. This process also calculates a host of other metrics to facilitate downstream analysis.

Note these three categories are focused on separate analysis goals or tasks of the gene spot quantification problem. In the GenePix [14,15] package (as shall be discussed in Section 2.4.4) these three categories are actually encapsulated in the manual operator identification stage and, therefore, the GenePix process sidesteps such issues.

Within each category there are many methods that have been developed that attempt to clarify the underlying surface in some way as accurately as possible. This section is not meant to be a concise review of all methods that exist in these categories; it aims to stress to the reader the problems faced and thus imparts a reasonable understanding of the concepts as they pertain to microarray analysis.

2.4.1 Addressing

Addressing is concerned with the general localization of the genes residing on a given microarray chip. Although the hardware itself (through operator settings) imparts this basic structure on to the slide during the spotting process, due to various real-world artifacts, the idealized positions do not in fact always hold true (as can be seen in Figure 2.6). A typical microarray slide consists of several blocks of gene spots. Each block contains several hundred genes, but critically these genes and the blocks move slightly across the surface. Matching an idealized model of gene spot positions to the actual scanned image involves the determination of several critical parameters. These parameters attempt to quantify the gene movement and can be split into three broad sub-categories as follows:

1) Separation - since a grid is an ordered matrix of gene spot positions, the distances between the columns and rows of the grids (as well as distances between individual gene spots) must either be supplied by a system operator or calculated from first principles.

2) Translation - gene spots within the grids are positioned onto the surface by the metal pins of the matrix structure. Due to everyday usage, these pins may suffer from slight irregularities (bent pins for example) in their positions in the matrix. Also, glass slide systems typically allow the operator to crop the scanner image (creating significant positional variation within a set of images). For multiple-color microarray images the so-called channels can also be misaligned with each other.

3) Rotation and Skew - ideally, rotation and skew should not be present for such systems; however, in practice minor issues are seen at master block and meta-block levels.

The problem of this so-called grid alignment issue has typically been addressed by one of two approaches in the past. The first approach focused on the hardware side of the equation, and proposed more accurate spotting technology. Affymetrix [2] is one such hardware solution to these alignment issues. By forcing building of the nucleotides onto the slide surface one nucleotide at a time, the photolithographic approach taken by Affymetrix guarantees finer tolerances than the cDNA hardware can at present. However, this greater accuracy generates greater expense. The Affymetrix system costs far more initially to setup for an experimental run than the cDNA approach. But, once this initial setup costs are completed, a major advantage of the Affymetrix approach over that of cDNA is that further experimental runs are cheaper to make (the Affymetrix template need only be created once).

The second approach to this accuracy issue then has focused more pragmatically on the software side of the equation. Here, the mentality goes along the lines of fixing the problems after they happen. Ideally, the hardware approach warrants greater time and expense as it ultimately provides better solutions. But as stated, with costing considerations of hardware solutions prohibitive for laboratories and such, the software approach is the option of choice at present. These software solutions then focus on template based and data-driven approaches to the alignment issues. Indeed, such template based solutions are the most prevalent approach taken by articles in the literature, what with

GenePix [14, 15], ScanAlyse [69] and GridOnArray [183], SPOT [52, 227] for example. Typically, these systems enable semi-automatic grid alignment by using a manual template matching process which allows the operator to adjust spot size, spacing and grid location from the outset. This is fine for perfectly regular grid structures, but problems arise for the irregular grid variety as can be seen in most of the template based solutions.

Data-driven software solutions are either based on 1-D surface projections as in SPOT, Steinfath [199], and Jain [111] or some other statistical approach as seen in Katzer [117] and Chen [44]. Various problems with these approaches are for example, gridding alignment based on axis projections of image intensities with problems materializing due to grid misalignments as in Jain. Steinfath make assumptions about the distributions of the images intensities and then allow distortions of the final grid positions. Such data-driven processes rely more critically on the underlying data quality mechanisms and have alignment issues with respect to missing gene spots and associated parameter sets.

Once the gene spots have been adequately addressed, a grid like structure is created such that either the outer boundaries of the individual genes within their respective blocks are demarcated or the gene spot's center positions are identified.

2.4.2   Segmentation

The next task is to look within the so-called grid rectangles and classify the pixels into two distinct groups, either pixels that make up a gene spot or pixels that represent the gene spot's local background. Segmentation can be defined as a partitioning of an image into regions depending on specific criteria [193]. In the case of microarrays, this partitioning ideally consists of two distinct groups; one representing the gene spot (foreground) and the other the background. Although it is the foreground area that is of interest it is important to quantify the background as well. As the targets bind to a pre-printed surface the bound foreground contribution is contaminated with non-specific probe. This non-specific probe has to be accounted for in some way. Normal practice is to subtract the local background from the foreground to render a more accurate gene spot. It should be noted however that at present there is no standard of what is considered an appropriate background region. Figures 2.6-2.8 highlight the non-conformity of this background separation issue by highlighting the gene spot background region as used by several commercial or academic packages.

The most popular background separation method is that associated with GenePix and can be seen in Figure 2.6. These diamond constructs located at the valley positions (center of a four gene spot region) define the background region metric of the central circular gene spot's foreground metric. The pixels falling in these circular or diamond regions have their mean or median values calculated such that they represent the intensity of the foreground or

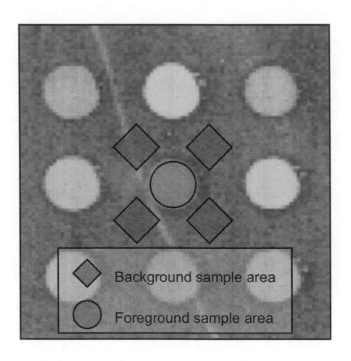

FIGURE 2.6
Gene spot background region using GenePix

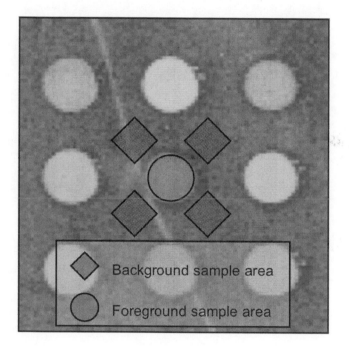

FIGURE 2.7

Gene Spot background region using ImaGene

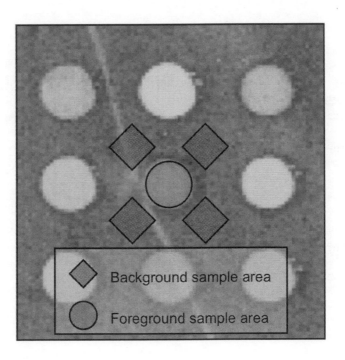

FIGURE 2.8
Gene Spot background region using ScanAlyze

background group respectively. As these valley constructs are central with respect to the local gene spots it is possible that this valley method's background calculation yields a measurement that is not representative of the actual background.

The circular background region of Figure 2.7 is an attempt to overcome such issues by taking a more localized measurement. An obvious problem with this background measurement approach would be that the gap between the gene spot and background region should be big enough. Too small a gap will mean that the background calculation includes pixels associated with the gene spots border. The border pixels of a gene spot consist of genetic material similar, but not actually gene spot material. Inclusion of such border pixels into the background calculation will render artificially high background estimates. Too large a gap on the other hand will either result in values similar to the valley approach (and the issues thereof) or start to impinge into neighboring gene spots.

Variations in a microarray images background can be quite intense, as such; Figure 2.8 is representative of a more global background calculation. In this case, the background regions themselves cover every pixel in the image except the gene spot pixels. Such a background region should be more representative of the true background distribution. A potential problem with all of these approaches however, is that the gene spots themselves may not infact be circular (ideally they should be, in practice however this is seldom the case). As such, more recent background calculation approaches have proposed the use of seeded region growing [1], watershed [208] or adaptive snake shape techniques [12, 224]. Indeed, segmentation techniques can be broadly classified into one of four groups, *Histogram or Threshold*, *Edge Detection*, *Region Growing* or *Clustering*. The remainder of this section presents a broad overview of these groups, along with a breakdown of some literature, be it directly relevant or related in some way to the segmentation of microarray images.

2.4.2.1 Histogram and threshold techniques

In histogram based segmentation, the histogram of the image is used to set various thresholds in order to partition the given image into distinct regions. It is expected that each region within the image will have some mean grey level intensity which represents a bimodal distribution. By examining the various peaks of the histogram the levels can be used as thresholds to partition the image. This powerful technique is spatially blind, thus illustrating that the two modes of the distribution may uniquely represent the gene spot and background regions. This assumption however could lead to incorrect observations. In some cases the images could be of such a poor quality that the pre-processing step might not improve the contrast between the elements enough to present two or more peaks for threshold selection. Unimodal distributions are typically obtained when the image in question consists mostly of large background areas with small regions of interest. Such techniques tend to lend themselves to the analysis of medical imagery and specifically microarray images.

In 1998 Cheriet [45] extended the work of Otsu [158] and presented a general recursive approach for image segmentation based on discriminant analysis. The approach segments the brightest homogeneous object from the image surface at each recursion. Essentially, the image's histogram is calculated for all iterations with the largest peak being separated from the others. This method does not imply any constraints on the initial number of objects in the image. This process continues until no more peaks are left in the histogram. As per Otsu's method (minimizing the weighted sum of within-class variances between the foreground and background domains), this technique has problems when there are more than two classes within the data. The authors concluded that the method gave good results when the target element represented the global minimum and is the darkest object, but failed in other situations.

Bhanu and Faugeras [27] proposed a gradient relaxation algorithm for solving unimodal problems and compared it to a non-linear probabilistic relaxation algorithm which was proposed by Ranada [172]. The overall process is based on the assignment of pixels to an initial probability of occurrence, which is then used with the gradient relaxation function to determine the relationship between a pixel and the eight immediate neighbors. This relaxation process is iterative in nature with the pixel values changed such that the histogram is no longer unimodal, this means a threshold value can be detected more appropriately.

2.4.2.2 Edge detection techniques

Edge based algorithms simply look for an abrupt change in intensity value where these individual pixels are connected in some immediate configuration. In practice, the change is not typically abrupt due to the optics of the capturing device and the sampling rate for example. Although more powerful than the fixed technique and able to cope with lower quality images, these edge methods have problems with irregular shaped gene spots.

Ahuja [9] used a pixel neighborhood method for image segmentation, where each pixel's neighbor was identified in a fixed size window. The aim was to generate a weighted matrix such that when multiplied with the appropriate vectors would yield a discriminate value that classified a pixel into one of two classes. They performed the experiment on 10 images, where two choices of a feature vector where chosen for every pixel. The authors concluded that from the obtained results, a pixel's grey level, along with those of its neighbors are a good set of features to use in classification.

Perkins [165] acknowledged that edge based segmentation has not been very successful primarily due to small gaps that allow merging of dissimilar regions. To avoid these problems, the author proposed an expansion/contraction technique where the edge region is expanded to encompass these small gaps and then contracted after the regions are labeled accordingly. Edges are then thinned with the result such that regions of differing intensity should be separated.

2.4.2.3 Region growing techniques

Region growing methods are perhaps the most powerful with respect to shape identification, but they rely on a seed growing methodology. The region growing algorithms take one or more pixels, called seeds, and grow the regions around them based upon some homogeneity criteria. If the adjoining pixels are similar to the seed, they are merged with them within a single region. The process continues until all the pixels within the addressed area are assigned to either the foreground of background groups. This seed selection process can be determined manually or automatically. For automatic selection, the relevant seed positions can be based on finding interesting pixels (brightest pixel, histogram peaks, etc.,) within the image. The manual method on the other hand relies on the user selecting these positions. A similar process would be the watershed technique which essentially operates in reverse.

Adams [1] studied the effectiveness of the seeded region growing (SRG) approach for greyscale images by using manual selection. Each of these seeded regions is connected and consists of one or more points with the border pixels calculated. The neighboring pixels are examined such that depending on whether they intersect any region from the initial set, the difference between a pixel and the intersected regions are computed. This new region becomes the input to the next iteration. A problem with the technique is that it is pixel order dependent, processing right-to-left top-to-bottom for example will produce a different result to processing in the opposite order. However, the authors concluded that the method is fairly robust, quick, and parameter free, but has an obvious problem due to its dependency on the pixel ordering process.

Mehnert and Jackway [140] improved the SRG algorithm by making it independent of pixel ordering. Their algorithm retains the advantages of its namesake, but if more than one pixel in the neighborhood region has a similar minimum measure, then all of them are processed in parallel. No pixel can be labeled and no region can be updated until all the other pixels (with the same status) have been processed. If a pixel cannot be labeled, (because it belongs to two or more regions for example), it is labeled as tied and further processing is stopped, at the end of standard processing these so called tied pixels are re-examined. The authors concluded that ISRG produced a consistent segmentation because it was not dependent on the order of pixel processing.

2.4.2.4 Clustering techniques

Clustering is a prominent unsupervised learning technique used for image segmentation that allows the search and grouping together of pixels with similar intensity values. The observed values are separated into distinct regions based on one of two basic properties of intensity: discontinuity and similarity. Various methods of clustering can be utilized, with the most common including [24, 118, 139].

Techniques such as k-means [139] and fuzzy c-means [66] are good representatives of the clustering paradigm and are still in widespread use today. k-means for example partitions data by initially creating k random cluster centers which define the boundaries for each partition in the hyper-volume. Each data point is then assigned to the nearest cluster center which is recomputed using their current members. The entire procedure is repeated until some convergence criterion is met such as; no reassignment of data points or a minimal decrease in squared error.

Fuzzy logic based algorithms [66], on the other hand, are specifically designed to give computer based systems the ability to account for the grey or fuzzy decision processes often seen in reality. Fuzzy logic based clustering therefore offers inherent advantages over non-fuzzy methods; as they cope better with problem spaces which do not have well defined boundaries. Essentially a membership function is assigned to each element which describes the degree of adherence to a particular cluster. These membership function values are in the range of zero to one, where the sum of each element's membership cannot be greater than unity. It is these membership functions that are minimized in order to alter the weighted difference and hence give the convergence property required for clustering. The weighting value also helps with the reduction of noise and other outlier type values that would otherwise bias the resulting analysis.

Like many techniques, the two clustering methods mentioned suffer from a variety of problems: they are heavily influenced by initial starting conditions, can become trapped in local minima, do not lend themselves to distributed processing methods and are restricted in their application to the memory resources of a single workstation.

Frigui and Krishnapuram [91] addressed three major issues associated with conventional partition clustering, which are sensitivity to initialization, determining the number of clusters and the sensitivity to noise and outliers. Their proposed method, robust competitive agglomeration (RCA), used a large number of clusters in order to help reduce the algorithm's initial sensitivity and determine the actual number of clusters by a process of competitive agglomeration. This method combined the advantages of hierarchical and partition based clustering techniques. Fuzzy membership methods where used to overcome overlapping clusters. Results proved that RCA could provide robust estimates of prototype parameters even when the clusters varied significantly in size and shape, and when the data was noisy.

Bozinov [32, 33] focused on applying these techniques directly to microarray image data and proposed an abstraction of the K-means technique whereby pre-defined centroids were chosen for both foreground and background groups. These pre-defined centroids were needed to overcome computational issues associated with clustering a full microarray image as mentioned earlier. However, the Bozinov approach biases the centroids towards the dataset's outlying values (saturated pixels for example), rather than the gene spots themselves. Remember that the reason for this segmentation process is to acquire a single representative measurement of each gene spot in the image. The more accurate this representative, the more accurate the biologist's interpretation of the data will be. The final stage then of a given microarray analysis process is that of gene spot value extraction. Note also that although not explicitly stated here, Markov models can and have been used as segmentation techniques as will be highlighted in Section 2.5.

2.4.3 Feature extraction

As the gene spot locations have been identified and the probable foreground and background regions determined the task becomes a one of numerical quantification. For every gene spot found in the microarray, a set of metrics should be calculated not only to aid in quality determination, but also to facilitate downstream processing. These metrics differ slightly depending on the analysis software used, but generally the following metrics are included.

1) Means and Medians - On microarray imagery used in this book, a gene's region of interest contains approximately 1000 pixels. Of this, the gene spots typically fall between 100~150 pixels, with the remainder being allocated to local background (as per the approach of images of Figures 2.6-2.8. The mean and median are used to quantify these pixel members into a representative measurement. This means for every gene spot in a two color microarray system, eight measures will be calculated. Simple quality criteria can be determined from the differences between the two measurements (as medians are more robust to outliers in a set).

2) Standard Deviations - The variability of a gene spot and its local background are also computed as they represent good measures of gene spot quality with respect to all of the genes in the surface having a similar value. Other measures are also used to improve confidence in the resultant regions being gene spots. A small variance change and high level of circularity between features for example.

3) Diameter - The diameter of the gene spot.

4) Pixels - The number of pixels that represent the gene spot and its local background region.

5) Expression Ratio - The log_2 ratio (of the gene spot designated median pixel values) is used to quantify a given gene spot's expression into a single representative value.

6) Flag - A flagging system is typically used to allow an operator to group suspect gene spots for later usage. For example, high quality gene spots could be marked with a flag value of 100, while poor quality ones have a

value of zero. In this way, it is easy to disregard poor quality gene spots from further analysis. However, care must be taken with such systems as quantification could be erroneous.

Of these gene spot metrics, the log_2 ratio is deemed the most important. This log value will be used in further downstream analysis by biologists to interpret the relationships between the experiments gene set. The log_2 calculation takes into account the given genes signal and noise (not gene signal) measurements from both channels (as determined by methods presented in Figures 2.6-2.8) of the image and can be extended from the simplified Equation (2.1) as follows:

$$\text{ExpressionRatio} = log_2 \left(\frac{\text{Median}(Cy5_{signal}) - \text{Median}(Cy5_{noise})}{\text{Median}(Cy3_{signal}) - \text{Median}(Cy3_{noise})} \right) \quad (2.1)$$

where *Signal* and *Noise* indicate the set of pixel intensities for a given gene spot.

At this time, these representative gene spot measurements are used to calculate a log_2 ratio such that the foreground and background elements of both image channels are used. This ratio uses all of the available channel information for accuracy purposes and the calculation reflects a single fold increase or decrease in the ratio of one channel to another with the same magnitude.

With the bringing together of these three categories (*addressing, segmentation*, and *feature extraction*) appropriately, a microarray image's biological characteristics will be rendered accordingly. In order to give some idea of how these three stages are usually applied in practice, the key stages of a commercially accepted microarray analysis package are briefly discussed in the next section.

2.4.4   GenePix interpretation

Various packages such as ScanAlyse, GenePix and QuantArray [14,15,69,169] are being used in both commercial and academic environments in an attempt to analyze microarray images as accurately as possible. These packages are designed with the common goals that are orientated towards finding the gene spots, separating them from their local background and quantifying their expression or intensity levels. For the purposes of result validation, one of these packages was used to allow comparison of a like nature to be drawn from a given image file. The GenePix package produced by Molecular Devices was chosen due to its broad usage and acceptance by molecular biologists. The package does not represent a *gold standard* as such, but rather an accepted methodology for microarray image analysis.

Figures 2.9-2.10 presents the GenePix methodology in the broadest terms to facilitate the understanding of addressing in this context. The goal of the addressing stage of analysis is to uniquely identify the areas of interest (the gene spots) within a microarray image. The gene spots are normally deposited

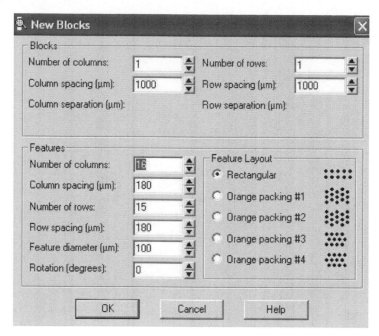

FIGURE 2.9
Parameter settings

onto the slide surface in groups proportional to the size of the printing matrix structure. The first stage of addressing therefore is to identify the location of these blocks to allow the later identification of the gene spots themselves.

Once the microarray image has been loaded into the GenePix package, the operator is presented with a *Parameter Settings* dialog window, as shown in the left image of Figure 2.10. The dialog window has to be filled in with appropriate values[††] for the internal algorithms to have a good chance of correctly aligning the blocking structures to the data. It is important to note that at this stage the values do not necessarily have to be 100 percent accurate, as long as they are close enough for the generation of the appropriate blocking template structure. The template structure must then be manually positioned over the image area as close as possible to the true block edges for the addressing algorithms to function correctly, as shown in the left image of Figure 2.10.

With this template structure set appropriately, the GenePix alignment algorithms can then be executed. These algorithms parse the underlying image surface and essentially perturb the template position within given tolerances

---

[††]Details such as the "number of columns," the "number of rows" for master and meta-blocks, the inter-block spacing, gene spot spacing and diameter are included in this dialog window.

FIGURE 2.10
Left: Master block identification; Right: Meta-block shape identification

while searching for some minimum error value. As long as any local artifacts do not fluctuate greatly and encompass the entire block area, these algorithms perform quite well, as shown in the left image of Figure 2.10. As a result of the addressing algorithms execution, the *master blocks* are adequately identified with the appropriate spacing and rotation values. Any differences between the initial parameter settings and these algorithmic determined ones are then updated accordingly. A close-up of the top right master block in the left image of Figure 2.10 is presented in the right image of Figure 2.10.

Generally, the addressing algorithms positioned the template structure appropriately. However, one obvious discrepancy is the misalignment [47] of several gene spots in the image. In certain cases the templates gene spot areas are not positioned accurately over the image due to misalignment or incorrect diameter settings. In these cases, it is left to the operator to manually realign and correct any diameter inaccuracies. The left image of Figure 2.10 has 9216 gene spots spread across its surface. Depending on the severity of the problems generated, the manual addressing process can take several hours to correct. Although the focus of the chapter is the overall composition of the microarray image, it is important at this stage to discuss issues associated with gene spot morphologies. The GenePix addressing process identifies the gene spot shapes at this stage due to the human steps of its implementation. As the framework has been designed to keep this human step to a minimum, the addressing of the gene spot locations and the identification of accurate shapes has been decoupled. This decoupling allows the addressing stage to focus on the task of gene spot identification, with a later stage determining the gene spots shapes themselves.

2.4.5 Gene morphology

The meta-block determination generally suffers from irregular gene spot morphology or significant local artifact noise. Due to the nature of the biological processes involved, the gene spots do not typically have perfect circular morphology. Figure 2.11 gives an example of GenePix gene spot identification, which highlights the differences that can be generated by the internal GenePix feature identification algorithms depending on the assumed underlying gene spot shape.

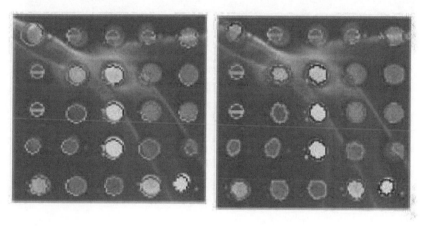

FIGURE 2.11
Left: circular morphology; Right: irregular morphology

The left image of Figure 2.11 presents the area when scanned with a circular detection method, which shows that the gene spots have been identified with circular masks. Pixels that fall within these circular areas would be classified as the gene spots signal, while the localized outlying region would be classified as background noise. It is important to note that the positioning and diameter determination of the gene spot template over most of the features in this image are incorrect.

The same area is also scanned with GenePix's irregular shape identification method as shown in the right image of Figure 2.11. This process generally performs better with respect to the gene spot diameter issue of the left image of Figure 2.11. However, both methods failed to identify several gene spots completely. These gene spots are individually flagged as "missing" or "not present" depending on some internal criteria or as determined by the operator.

The final stages of processing in GenePix revolve around creating a set of summary statistics for each gene spot found. However, as mentioned in Section 2.4, there is no standard definition of a gene spot's local background should be. As such, there will be variability with a given gene spot's measure depending on the local background and morphology methods used. Examples of this variability can be seen in the background enhanced images of Figure 2.12. In each image, the left-hand part represents the original image surface, while the right-hand part highlights the surface variation within the left-hand surface.

FIGURE 2.12

Examples of slide surface structure and signal variation. Left: the good; Middle: the bad; Right: ugly and co.

The left image in Figure 2.12 illustrates what could be classified as a relatively "good" microarray image. The background fluctuates slightly, but generally remains relatively uniform. Although variation in the background will have an effect on any measurement, in this case the impact of this background variability should be minimal. The surface of the middle image in Figure 2.12 represents a relatively "bad" surface where the background variation is high and as such any raw measurements taken will be compromised. The right image in Figure 2.12 presents two extreme examples of what is possible within a microarray image; on the left the surface area is littered with areas of significant background variation. Measurements taken in these areas will be adversely affected. Also, this image has significant Cy3 (green) channel noise which means that the two-channel data for this image is not balanced and this is possibly due to poor dye preparation. The image on the right looks similar to that in the middle, the significant difference is this image has a large amount of negatively expressed genes. This means the local background of these gene spots will be of greater intensity than the gene spots themselves. As shall be discussed later, this reversal of intensity relationship between the gene spot and its local background can cause great difficulty for algorithms.

2.5 Copasetic microarray analysis framework overview

The potential knowledge to be offered by microarray technology to our current understanding of biological processes is vast. However, as hinted at by various computer vision challenges as discussed in the previous sections, such technological prowess is still in the early stages of development. In order to truly exploit these systems, not only will current throughput capabilities need to be extended significantly, but the knowledge gained from the image analysis processes themselves will have to be harnessed to a greater extent. One way in which such knowledge can be improved is to remove the human from the image analysis process completely. The creation, therefore, of such automatic analysis systems improves the result repeatability of such processes. Before briefly describing the key stages of the proposed CMA framework, a review of contemporary systems in the literature that would seem to have similar goals to the CMA approach is discussed.

Initially a system developed by Jain [111] was proposed as an automated method for the processing of microarray images. Rather than require a human operator to input coordinates of the image, the algorithm used the geometry of the slide (the block layout) to create an appropriate template structure. Based on minimization criteria which compared the horizontal and vertical profiles of the image, the process was able to align and resize the image such that features were aligned with a predetermined edge. Failure of this process in noisy or badly hybridized slides results in operator intervention.

Angulo [11] produced an algorithm that attempted to process a slide automatically, based only on the gene spot characteristics. In principle, a series of dilation and erosion image operators could be applied to the image in order to detect the features of interest. This technique when combined with the use of a simple threshold process was shown to be sufficient to extrapolate the gene blocks of an image. A similar process was then used to detect the gene spots within each block. Although successful on clean images, this process quickly degrades when high intensity or large artifacts are present in the image surface.

Although methods as proposed by Jain and Angulo were showing promise, they were unable to deal with full-scale microarray images. Work therefore continued in these areas and systems were introduced that were designed with the aim of processing full slide images automatically [32,33]. Bozinov focused on a web based approach making paramount the accurate and automatic detection processes as the operator will only play a limited role in the analysis process. Using an adaptation of the k-means clustering method [139] a binary image of the microarray was created. This binary image could then be used to calculate horizontal and vertical profiles of the image to aid in the detection of the centroids of each gene row and column. Unfortunately, there are no details available for these stages of the process other than graphical charts. It is difficult therefore to comment directly on the accuracy of such techniques. One observation that can be made, however, is that all of the images used in the aforementioned papers have very low levels of background noise with the majority of gene spots present.

Other papers present ideas along similar themes with the application of wavelets and Markov random fields [16,100,115–117,209,210] as with all of the previous examples, however, these papers do not use large-scale experimental data. More often than not the imagery they use is taken from cropped sections of well-hybridized, "clean" slides. What is required is a highly automated system that is designed to work with poor quality image data rather than the cherry picked images of previous techniques.

The proposed framework system - *Copasetic Microarray Analysis* (CMA) - takes the various process challenges as discussed throughout the chapter and breaks their concepts up into logical components. The core events of the logical components were formulated further into design requirements. The requirements addressed specific limitations or weaknesses of existing microarray analysis techniques. The greatest such driving limitation of current techniques is the need for human intervention at the very beginning of the analysis stage.

Recall that on completion of the digitization stage, a gene chip or slide of the resultant microarray experiment is loaded into an analysis package. The biological technician must then manually demarcate all of the gene spots within the appropriate image. Although such a manual process is typically augmented by various semi-automatic stages, as the nature of these semi-automatic processes is to apply more weighting to the technician derived gene spot location, the internal spot finding algorithms can generate inaccurate locations. The technician is then required to spend more time with refining the algorithmic solutions. Indeed, if the technician were to accurately demarcate all of the gene spots manually (does not use the spot finding algorithms), not only would more time be spent on this demarcation task, but the repeatability of such a demarcation result would be dubious at best.

The question then is how can the technician be removed from the analysis stage without reducing the quality of the resultant analysis process. Ultimately, the technician should be completely removed from the analysis task, in practice however, this is infeasible with current microarray systems (biological protocols and microarray systems are simply too immature). Therefore, the question should be rephrased to, how the technician's time requirement be minimized without reducing the quality of the resultant analysis process.

Clearly, the majority of the technician's time is spent in the demarcation process as mentioned above. In order to overcome this requirement, appropriate functionality must complete the technician's demarcation task. However, the technician is used primarily as the microarray images are littered with significant background noise which make difficult the separation of gene spot and artifact (something the human vision system is very good at). Removing the technician, even minimally, has knock on effects however. During the demarcation process, the technician is not only identifying the gene spots, but they are ideally determining the true shape and position of the spot. Such processes are non-trivial in realistic situations, as can be attested to by the poor repeatability of a given analysis. A major benefit of proposing this fully automatic system then, not only includes minimizing the technician's time, but also increasing the repeatability aspect of a given image analysis.

Founded on this premise of time reduction and result repeatability then, a fully working truly automatic analysis system would be beneficial to many laboratories. As such, a set of modules are developed which facilitate the blind processing of an entire microarray experiment. From conception, these modules were designed such that they should be goal orientated. Although the components are executed in a sequential manner, there should be no reliance on any single module being entirely accurate. Rather the modules should have some form of redundancy built-in such that they cover each other slightly. This means the modules are designed to work together in a collective manner, such that errors generated by one module will typically be rectified by another. Such built-in redundancy ability has the knock on effect of facilitating the processing of images that are significantly worse than those tackled by techniques currently presented in the literature.

Unlike other automated techniques that have been proposed to this affect, the CMA framework is not rigid. It is the framework's modularity based construction which renders its adaptability responsible for robustness. In Figure 2.13, a skeletal structure of this framework is presented, showing the key stages from the original input image through to calculating the various output metrics.

Each of the stages in Figure 2.13 are goal orientated, which means that techniques can be "swapped out" to allow various computational tasks to be conducted should the need arise.

The first stage of the proposed CMA system would be to clean up or reduce as much of the background artifacts in an image as possible, whilst leaving the gene spot intensities as they are. Given that the scanner devices used in the digitization stage have a maximum detection level of 65535 unique values, an ideal situation would have the gene spots residing in the low-to-upper middle intensity range. Gene spots with intensity at or near the maximum threshold would be saturated; this means their true intensity level is unknown. Genes spots at or near the other end of this scale (zero) would be indistinguishable from the local background. For gene spot identification purposes then, the image data can be rescaled into the range of 0-255. If accomplished correctly, this rescaling will not only retain the majority of gene spots in the higher range,

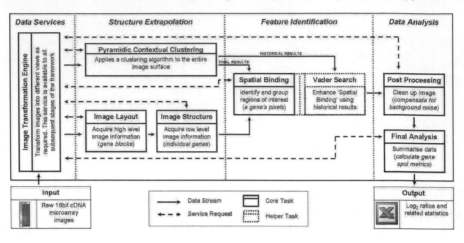

FIGURE 2.13

Copasetic microarray analysis workflow diagram

but will also enhance those in the lower range of the scale. As shall be seen in the novel process of Chapter 3, this concept of rescaling can be extended to enhance various parts of an image such that the framework's algorithms have more knowledge of the underlying image structures. The proposed *Image Transformation Engine* (ITE) of Figure 2.13 is the only component which has direct access to the raw microarray image data. The purpose of the ITE is to supply this data in modified and raw forms to the framework's components as requested. For example, if the quality assessment process determines that the data is lacking detail in a given area, a low-pass filter could be executed to improve said area's analysis.

With such a reduction process complete, the task at hand then becomes one of gene spot identification. Although a relatively trivial task for the technician's vision system, the computer vision task is a little more complicated. Essentially, the computer will be presented with an image full of intensities at this stage. However, whereas the human vision system sub-consciously looks for pattern in the data using a multitude of subtle hints, the computer's vision system is not so intelligent. Therefore some spatial element should be taken into account when examining the individual intensities. With such a spatial orientated process, the consistent underlying pattern of gene spots should become emphasized (in some way) more than those of the inconsistent background intensities. In Chapter 4 it shall be shown that this enhanced computer vision process is possible with the creation of a novel spatial clustering technique. The *Pyramidic Contextual Clustering* (PCC) method (as seen in Figure 2.13) facilitates the requirement of full-slide segmentation. The component allows the execution of traditional clustering methods to be applied to very large image datasets. This technique also reduces the processing and memory requirements to those feasible with modern desktop computing technology. With such a data reduction and spatial awareness process in place, the identification and separation of the individual gene spots should become easier to achieve.

Although the spatial awareness process results will more than likely contain missing gene spots due to intense background noise and other such anomalies, these results should be sufficient to guide a gene spot identification task. At this stage of the process the identification task does not need to calculate the appropriate morphology of the individual gene spots but rather, determine the compositional aspects of the underlying image as a whole. Chapter 5 details such a structural mapping process that can identify the master grid regions the gene spots were spotted to initially as well as the individual gene spot locations. The *Image Layout* (IL) component uses a variety of techniques to ascertain the general layout of the image surface. With these areas of interest defined, *Image Structure* (IS) then uses similar processes albeit at finer granularity in order to discover the detailed structures of the surface. These compositional stages (see Figure 2.13) can either utilize the raw data or more beneficially, one of the alternate views as provided by the ITE or PCC services.

With a given microarray image underlying grid structure thus identified, the gene spot identification task can commence. Such an identification task should produce results similar to those as presented in the gene spot overlays of Figures 2.6-2.8. Chapter 6 will show how the above mentioned stage results are brought together to determine the gene spot morphologies as accurately as possible. Indeed, if the spatial awareness process is designed correctly, insight into the gene spots appropriate morphology would be forthcoming. Gene spot morphology calculation is achieved by the *Spatial Binding* (SB) component and the *Vader Search* (VS) helper function (shown in Figure 2.13), which can use estimated gene center positions and PCC results, to search for and combine groups of pixels that meet some criteria. This process can be completed for the genes that were well defined and the information gained used to rebuild missing genes.

The final stage of such a process would quantify the aforementioned gene spot regions into their representative log_2 measurements. The calculation stage would also generate appropriate statistics for the underlying gene spots such that downstream analysis can be carried out on the data. Using the structural information identified and the raw image data, the genes on the microarray are analyzed. Generally this analysis consists of two phases; one of *Post Processing* (such as background correction) and a second of *Final Analysis*, (e.g. consolidating genes pixels into one representative value), see Figure 2.13.

2.6 Summary

This chapter provided an overview of the microarray research domain. Initially a review of the genetics field was introduced. The review was designed to give the reader an appreciation of key genetics concepts that pertain to the book subject. Then, a detailed review of microarray technologies was presented. This second review was designed such that the reader would not only appreciate the challenges involved in the relatively new field of microarray image analysis, but crucially, would gain a basic understanding of the core concepts and techniques used within the field. In effect, the review elucidates indirectly how microarray technology harnesses the central dogma concept of molecular genetics and in so doing revolutionizes modern day biological research opportunities.

The challenges involved in correctly analyzing a microarray image were discussed along with the deficiencies of current microarray analysis systems. Two major issues associated with current microarray analysis systems were found to be related to the operators of said systems. Whether the system utilizes guide spots or not, some form of operator process is required, first to correctly identify and demarcate an image's underlying gene spots. The second issue is more subtle, although the operator does a splendid job at identifying and marking the gene spot boundaries, they will inherently make different decisions when presented the same information. Although this could be argued as a trivial artifact of the process, the ability to generate consistent results for a given image is paramount.

For images containing thousands of gene spots (as is usually the case), this operator not only becomes a bottleneck to high-throughput analysis, but their inconsistent decisions render the ability to compare multiple image results more ambiguous. An appropriate process therefore was created such that the deficiencies of microarray analysis systems were identified and broken up into their constituent parts; the challenges associated with these non-trivial deficiencies were further identified and examined such that a reasonable attempt could be made on their rectification. The proposed *Copasetic Microarray Analysis* (CMA) framework has been designed such that these deficiencies can be overcome for the most part (human intervention has been minimized but not removed in the case of extremely noisy image data). As shall be shown in the forthcoming chapters, this fully automatic blind microarray analysis system performs extremely well when presented with a range of "clean" and "noisy" real-world microarray data.

With the problems as detailed above in mind and the proposed framework components formulated, the next chapter explores ways in which a microarray image can have various regions of its surface enhanced. Such an enhancement process is required if the realization of a fully automatic blind microarray processing system is to be achieved.

3

Data Services

3.1 Introduction

Chapter 2 discussed the biological concepts that microarrays exploit for their usage. These biological mechanisms were explained in broad terms along with a detailed description of their technological manifestations such that an understanding of the basics of complementary deoxyribonucleic acid (cDNA) microarray systems is clarified. As discussed in Section 2.5, our proposed *Co-pasetic Microarray Analysis* (CMA) framework attempts to operate in a truly blind manner. This means (excluding the failsafe measures of the *Image Layout* (IL) and *Image Structure* (IS) components) the framework operates without the need of manual assistance. The framework should perform the analysis tasks such that they are comparable to the GenePix type process as discussed in Section 2.4. However, due to the high signal variability seen across the test dataset (when working directly with the raw microarray imagery), identifying gene spots can be counterproductive at this early stage. Therefore, rather than use the raw image, it is suggested that producing views of the image data such that emphasis is placed on certain frequencies or regions of interest would not only be advantageous, but more effective in terms of the overall goal.

This multi-view analysis process is managed by the application of the *Image Transformation Engine* (ITE) component to the microarray image data. In the current implementation, the ITE component generates an image designed to enhance the positions of the gene spots. Although this positional information is partial in some areas of the newly generated image surface, it is fit for identification purposes at this stage.

As was highlighted in Chapter 2 the CMA framework's processes are designed such that the requirements of manual intervention are minimized. This design constraint creates a non-trivial search problem however. The microarray images are embedded with gene spots for the particular experimental run as required. However, due to the nature of the biological process and the desired caveat of no prior domain knowledge, the gene spot positions must be determined by an algorithmic method. In order for the gene spot addressing stage to return appropriate results, the raw microarray image should first be pre-processed to remove as much artifact noise as possible. The noise removal process should function such that the gene spots in the image are clarified while the background noise and other artifacts are either left as they are, or reduced in severity. Once this so-called enhancement process is complete, later stages of the framework can use the new surface information to begin the process of building up a more accurate picture of the gene spot locations.

The aim of the chapter is to explain how the proposed enhancement process functions. In the context of the framework as highlighted in Section 2.5, this process of gene spot enhancement is encapsulated in the ITE component as seen in Figure 3.1.

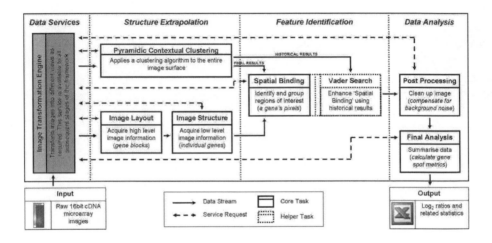

FIGURE 3.1

The CMA framework

The enhancement process will take the form of multiple views of the microarray image data with emphasis placed on the different intensity distributions thereof. This highlights two important objectives for the investigation.

1) The proposed solution should not lose any image information during the conversion process; practically however, lost information should be minimized with the majority of the gene spot regions left intact.

2) The process should execute in a reasonable amount of time with no prior domain specific knowledge (gene spot positions) required at input.

Based on these two objectives, Section 3.2 presents a broad overview of the proposed solution to this image enhancement process. The problems faced by the ITE component are first highlighted by a brief discussion into the surface variance challenges seen over a selection of test images. With the types of noise broadly highlighted, the discussion then details the preliminary research carried out for an appropriate surface enhancement stage. The section then discusses the algorithmic methods of the proposed ITE solution and the advantages of using a multi-view approach to microarray image analysis. The ITE process is then evaluated in Section 3.3 against a selection of real-world microarray data. The chapter concludes in Section 3.4 with a summary of the issues discussed and the solutions presented to enhance the imagery.

3.2   Image transformation engine

Irrespective of the type of data that is to be analyzed, it is normal practice to pre-process the data such that artifacts of one type or another are reduced in severity. In the case of a time series dataset, these noise objects could be incorrectly interpreted as valid data points for example. Pre-processing stages therefore are designed to reduce such interpretation issues. As discussed in Section 2.3, a microarray image is littered with various amounts of artifact noise that could cause an analysis algorithm to make erroneous decisions. Investigations into appropriate pre-processing stages therefore should focus on artifact removal in the microarray image surface.

3.2.1   Surface artifacts

Although microarray technology has been engineered to fine tolerances, there exist several areas of weakness where contamination can corrupt a required gene signal. Below, these weak areas are broadly detailed to give the reader a greater appreciation of the difficulties involved with the identification and segmentation of a microarray's embedded gene spots.

3.2.1.1   Biological contamination

The most obvious place for noise to enter this particular process is within the biological stage itself. Biology, as a natural process, relies on the biochemical interactions of the underlying four-base nucleotide components. A mutation or corruption of a single nucleotide within a sequence interaction can have a significant effect on the final hybridization results. There are many biological sub-processes that should be taken into account at this stage, but generally contamination in this context can be split into two groups.

3.2.1.2   Hybridization noise

As hybridization is a complex process, it can be affected by many conditions including but not limited too, temperature, humidity, salt and formamide concentration as well as target volume solution. One of the underlying problems with the hybridization step is the simple fact that genes are related to each other through the mechanics of evolution. Genes evolved from common ancestors with the result being the majority of them formed sets of closely related families and thus share sequence similarity. It is not uncommon therefore for cross-hybridization to occur between different genes during the hybridization process. Such cross-hybridization can manifest itself into the resultant image surface as excessive bounding of genetic material or erroneous pairings between sample and reference material (Figure 3.2 highlights some of the more common biological noise issues involving evolutionary processes).

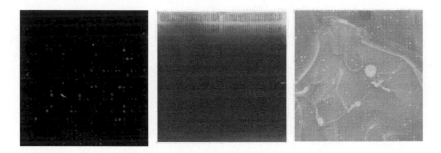

FIGURE 3.2
Left: weak or low intensity gene spots; Middle: hybridization solution dehydrated onto slide surface; Right: washing solution dried on slide surface

3.2.1.3   Washing noise

Once the hybridization process is complete, the slides are washed for two important reasons. First, due to the process being one of a biological nature not all of the cDNA within an individual experiment will have bound to the probe's DNA. This excess material must be removed to help guarantee that the final gene expression quantity calculated represents as accurately as possible the bound genetic material. The second reason is due to the hybridization process itself. As stated above, an affect known as cross-hybridization can occur in similar species whereby cDNA probes bind to inappropriate target regions. The washing process helps remove these weakly bound probes from the microarray surface thus increasing the stringency of the resultant data.

Figure 3.2 highlights some of the common biologically generated noise artifacts that can be generated during such evolutionary mechanisms.

The situation in the left image of Figure 3.2 is normally associated with the *polymerase chain reaction* (PCR) process not generating sufficient raw messenger ribonucleic acid (mRNA) material for the binding stage. The situation can also transpire as a result of generating poor quality mRNA genetic material (short sequence length for example). In the middle image of Figure 3.2 the pre-printed slide was not submerged into the sample genetic material correctly (or evenly), resulting in the target material binding to the majority of the slides surface. Such a result can also transpire when the hybridization process itself is left too long. The right image of Figure 3.2 is the result of poor wash protocols, generally this means the washing material itself has dried onto the slide surface (be it due to wash time or excessive wash material). The majority of these noise generating issues can be resolved by sound biological protocol and process, and should not be seen on a regular basis.

3.2.1.4 Systematic contamination

Another type of noise that can enter the process is that known as systematic noise. This type of noise typically has some form of internal order or structure and in the case of microarrays is usually a result of poor slide preparation and manual process. Importantly, note that technical repeats (control spots for example) may well be systematic in nature, but they are not mentioned as they are meant to be signal rather than noise objects.

3.2.1.5 Hardware noise

The biggest contributor to systematic contamination within the microarray domain is that of the hardware itself. A microarray device (as shown in Figure 2.2 consists of a robot arm that traverses a bed of slides. The matrix print-head structure at the end of this arm deposits the genetic material to pre-determined locations via the use of specially designed metal pins. A great deal of design acumen has been rendered into the production of the microarray device but the delivery mechanism (notably the print-head structure) does have problems. The images in Figures 3.3-3.4 highlight some of the issues caused by such hardware mechanisms. The left image in Figure 3.3 shows a common problem with microarray imagery. Here the gene spot morphology has in several locations become corrupted. There are more variables involved in these morphological issues than the hardware, but hardware design plays a critical role in their development. As highlighted in Appendix A, the print-head matrix of the microarray device is pre-configured, such that the deposited gene spots should be identical. Crucially, however, the print-head pins become misshaped during their normal usage, with such physical anomalies leading to gene spot shapes that are not perfect. Note that, ideally, the print-head should deposit "perfectly circular" gene spots onto the slide but, due to the anomalies as mentioned above and reality of the environment, this is rarely the case practically.

The right image in Figure 3.3 shows what can happen when a print-head's pin device itself does not take a consistent amount of the target sample from an appropriate well (due to physical damage perhaps). Also related to the morphology issue, the pins can deposit more material than desired that can lead to artifacts such as comet tails (seen here), donuts (so-called as there is

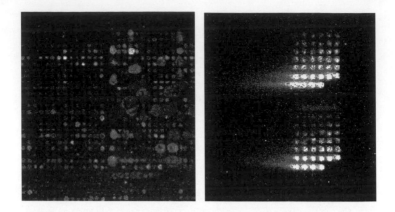

FIGURE 3.3

Common systematic noise issues involving evolutionary processes. Left: irregular spot morphology; Right: excess cDNA bound to slide artifacts

FIGURE 3.4

Common systematic noise issues involving evolutionary processes. Left: external gene block misalignment; Right: internal gene block misalignment (meta)

no genetic material at the center of the gene spot), black holes,* and bleached genes. The images in Figure 3.4 highlight misalignment issues associated with the pin shape and a microarray devices physical placement in both the master

---

*Not so much due to lack of genetic material but to the presence of spotted material that masks specific binding sites on the glass surface.

block (left image in Figure 3.4) and meta-block locations (right image in Figure 3.4) to the gene blocks respectively.

3.2.1.6 Random noise

Since the microarray process is biological in nature, random perturbations within an experiment should also be expected. As well as the nature of the imperfect control mechanism used in current analysis systems (humans) various external artifacts will typically appear within an experiment run. These artifacts include but are not limited to; hair, scratches, dust, and finger prints.

3.2.2 ITE precursor

The potential signal contamination issues discussed above represent the major sources of image contamination found in a typical microarray experiment. In order to overcome these and other noise related issues a systematic approach should be followed when analyzing the image data. What is needed at this stage therefore are processes that can reduce this artifact noise while at the same time enhance the signal of interest (the gene spots). Such dual functionality is achievable with a combination of data smoothing and intensity re-mapping techniques. Also, as a result of the re-mapping phase, memory load and processing time will improve significantly. This resource improvement is made possible as the raw microarray image data is stored in a 16-bit (2^{16}) format, with a large range of these 16-bits devoted to representing noise objects. The internal processing stages of the framework function directly on the ITE reduced image data (8-bits (2^8) format) versions of the raw data. The final stage component of the framework *Final Analysis* (FA) uses the knowledge gathered from these 8-bit images to determine the transcript abundance of the gene spot regions in the raw imagery directly.

However, due to the nature of microarray imagery, one technique will not be adequate in expressing the full range of issues seen in an image. Therefore a multi-view process must harness the appropriate information from several such views of the image and present a relatively "good" generalized result image. The problem itself at this stage consists of two related issues; first, the interesting or especially noisy regions of an image need to be highlighted, and second, analytical techniques which are appropriate for image enhancement need to be identified and applied to these interesting regions. In the case of the first issue, the regions of interest are the gene spots themselves. However, due to the lack of manual intervention at this stage in the framework, it will be unclear as to what a gene spot actually is. The second issue could use one of many analytical techniques for appropriate data transformation; the remainder of this section aims to gain insight into the techniques that are most appropriate for the ITE stage.

As stated above, the intensities associated with a particular region within an image channel can be quite different in power. Therefore, care must be taken with the conversion of the raw image data as these channel biases could be such that they cause subtle problems at later analytical stages. Note that traditional packages designed to analyze microarray images (GenePix for example) provide facilities for the operator to manually adjust the brightness and contrast of the image (and therefore circumnavigate this conversion process). This manual ability helps the operator to balance the intensities between two or more channels and thus make better approximations of a gene spots outer region albeit at the cost of the operator determining the appropriate settings for every gene spot in the image (in practice, one setting is chosen such that most of the gene spots can be seen).

Section 2.5 highlighted, and Figure 3.1 re-emphasizes, that the only required input into the CMA framework is the raw microarray image itself. It is the responsibility of the ITE to manage and control this raw image data, supplying it in the unaltered form and a variety of transformed alternatives. In this way, the framework's components will not be restricted to one view of the data as is usually the case, but will benefit from these multiple perspectives. Ultimately, this multi-view approach will also facilitate greater adaptability whereby the system does not simply succeed or fail at a given task, but can re-analyze a given region from a different view.

Figure 3.5 presents the final results of an ITE precursor study. This initial study was designed to identify and implement transformation techniques that would be useful to the ITE clarification process. As mentioned above, such clarification is needed for a fully automatic analysis process to deal with a range of possible input image anomalies. Without such an enhancement process, the internal algorithms will typically fail to achieve good results at later stages of the framework. Many such techniques were examined but four fairly simple transformation processes stood out as representing a good compromise between the transformed image and a rapid one-pass process. These four transforms are

1) Linear - provides a simple remapping of the raw input data's two or four channels from the 2^{16} range into 2^8. Such a scaling process renders an output image that is similar to the original raw image (as per a GenePix process) with the advantage that the resultant memory load process has been reduced.

2) Square Root - is a good compromise between the data reduction of the linear transform and the enhancement aspects of a remapping process. The square root transform places an emphasis on the low-mid-intensity regions of the raw input image. Such an emphasis results in greater gene spot clarity with the negative cost of increasing the low lying background region intensities at the same time.

3) Inverse - places greater emphasis onto the low range intensities which results in an output image that captures more background detail than gene spots.

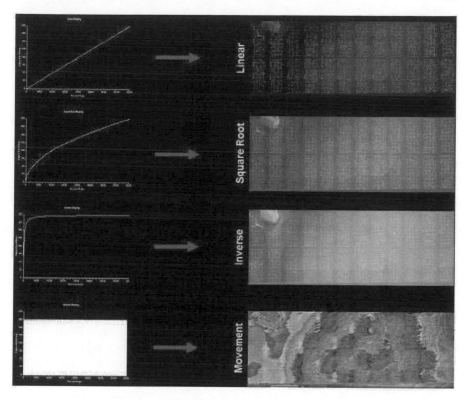

FIGURE 3.5

Summary of ITE precursor transforms

4) Movement - was designed to provide an understanding of the fluctuation or variability that can exist across the whole microarray surface. Regions of this surface that have more consistent movement (significant bits) will result in a remapping process that separates the gene spots from the background artifacts cleanly. On the other hand, regions with high movement will result in the separation been the two domains being greatly reduced.

The four transform functions are mentioned here for consistency, while Appendix B provides implementation details. Figure 3.5 presents the transform output images for a sample microarray image as processed by the four transformation methods. The input image was chosen as it represents a good middle ground for understanding the problems faced by the framework's algorithms. The image is generally "clean," with no extreme levels of noise throughout. The two artifacts in the image represent the kind of morphological and background intensity range challenges of general noise elements. The artifact region smothers a selection of the gene spots with an intensity range similar to gene spots in other regions, while the background of the image has an obvious bias (it should really be structure-less with a consistently low intensity surface). These data anomalies present different challenges to the various algorithmic processes. The graph plots on the left represent the frequency response curve for the particular re-mapping process, while the surface plots on the right present the re-mapped image.

These multi-view perspectives take a step in the right direction of image explanation as they emphasize different intensity based aspects of the image surface. On initial inspection of the figure's right hand images, one might suggest the linear transform as providing the greatest gene spot information. Closer inspection of the image, however, reveals that not only are gene spot regions missing from the image, but, there are large regions containing low intensity information. Note also that there is in fact no perfect solution to this missing gene spot problem (the best one can hope for is at least an indication of all the gene spots within the image). The default presentation of an image loaded into the GenePix package for example is very similar to this linear surface.

Importantly, however, this transformed image surface does not tell the full story, as there are significant noise objects (and gene spots) contained within the black regions of the image. The *square root* transform was thus designed to focus on the low-to-mid-intensity ranges of the raw image with the expected outcome clarifying these dark regions (as well as the whole image) at the expense of capturing more background noise. This transform has worked very well as the full slide surface can be seen clearly, with the artifacts as suggested by the linear transform emphasized to a greater extent. The gene spot regions have also become re-emphasized across the whole image surface to a greater extent than those of the linear process. Such a square root transform raises the question of what does the background region look like? The artifact in the upper left and the central region would suggest there is more going on in this surface than first realized.

The inverse transform attempts to answer this posed question by weighting the low-range intensity values more so than the upper ranges. Here, it was expected that the resultant transformed image would give a much clearer idea of the background noise present in the image than the raw linear image suggested. Although some definition has been lost from the region of the square root result, the artifact and indeed the transition zone of the blending region have been defined cleanly.

The culmination of the above transforms in turn raises a further question of how the surface changes. Such a pixel movement transform would yield insight into the consistency of the surface's intensities as the movement essentially retains the images most significant bits. Ideally, it would be expected that only the gene spots themselves would show extreme variability. In practice however, as indicated by the movement surface in Figure 3.5, the majority of the surface in this case fluctuates rapidly. With such intensity differentials in mind and a greater understanding of the potential data anomalies present in typical microarray imagery the ITE functionality can now be explored.

3.2.3 The method

As discussed in Section 2.3, a typical microarray image contains several kinds of artifact noise (hair, scratches and fingerprints for example). If the operator of traditional microarray image analysis software is to be removed from the analysis process, the input images must be cleaned (have noise artifacts removed). However, this cleaning process should not affect the gene spot intensities themselves, as later stages will use such information to help determine the gene spot locations.

Applying a smoothing operator before the actual re-scaling process takes place will be of additional benefit as this smoothing process can remove small region high intensity pixels that are of no value in a gene spot identification context. Due to this decoupling process the transform functions also become independent components that can be called upon by the ITE core as and when needed. Figure 3.6 presents this decoupled work flow process as applied to a dual channel microarray image.

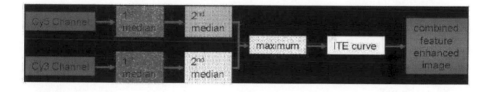

FIGURE 3.6
Pipeline of ITE feature response curve generation

From Figure 3.6 it can be seen that the two channels of the input image (in this case) are first smoothed by two different median filters before the actual ITE filter itself is applied. The first median filter simply parses the red and green channel surface's independently with a sampling window of 5×5 pixels centered on each pixel in turn. This center pixel is thus calculated as the median of all the pixels in the 5×5 region. The 5×5 region of the median filter process effectively removes any local high-intensity pixels with the minimum amount of disruption passed on to the surrounding regions. The second median filter is essentially the same as the first in that the second filter uses a larger window region and sampling ratio. In this case, the window measures 57×57 pixels (slightly bigger than a gene spot's local region) with the sampling or centered pixel set to every forth pixel in turn. The 57×57 pixel window size is calculated from an auto-correlative process derived from the work of Chapter 5. This second filter sampling process results in a simple estimation of the images background features[†] (if the median value is subtracted from every pixel in the image). Such smoothing operators had two positive effects on the image data: First, low-level background noise was either reduced substantially or removed altogether from the image. Second, large scale artifacts (such as that seen in the imagery of Figure 3.5) have their internal structure removed from the image data (thus creating a halo type effect).

With the two median operators removing "salt and pepper" type artifacts and reducing large artifact regions from the image respectively, the two channels are merged into a single combined view to help with the reduction of computational complexity. Testing resulted in the identification of the most effective and least costly method of merging the two or more images to be that of taking the maximum intensity value for a given pixel from the channels. Note that such median transforms as based on sub/superset kernels raise the question of a multi-resolution transform process being of great advantage to image enhancement.

From the square root image of Figure 3.5, one could argue that this transform contains enough gene spot information for the ITE function. Indeed, this transform was chosen initially, as it seemed to offer the fundamentals (the majority of gene spots in the image and good low-level background information). However, although the resultant output via such a square root transform does produce a relatively "clean" image, a lot of low lying (background similar) gene spots are lost in the process. If sufficiently large enough numbers of such gene spots are lost in the re-scaling process, later identification stages will have obvious difficulties. Also, as one of the reasons behind such a transform process is to enhance the gene spot regions, "lost" gene spots are non-desirable. It would make sense then to take elements from both the square root and linear transforms of Figure 3.5 as these two transforms contained highly-desirable surface features.

After various testing stages were carried out to determine relative performance and speed of execution, a good compromise for the ITE function was indeed found to be elements of the square root and inverse transforms. Further testing processes started to gain a clearer understanding of a surface's peculiarities, with an appreciation of the appropriate weighting levels needed between the two functions respectively. Ideally, such a hybrid function needs to harness the gene spot intensity ranges as calculated by the square root function while at the same time taking a higher percentage of the gene spot similar background intensities from the inverse function (the gene spot intensities are further refined at later stages). Essentially, the more an image's gene

---

[†]Over twenty million pixels in an image, with 9216 gene spots consisting of ∼120 pixels each, means just under nine million pixels are associated with the background.

spot regions can be separated from their local background at this stage, the greater the accuracy of the resultant image analysis process.

Although unlikely to be truly optimal for this ITE process, the current transform has proved to be highly efficient at preserving the required level of detail for image processing techniques at later stages of the framework. These ITE processes are summarized in Equations (3.1)-(3.3) respectively.

$$S(x) = \sqrt{x} \tag{3.1}$$

$$I(x) = 1 - \left(\frac{1}{\frac{x}{2^8} + 1}\right) 2^8 \tag{3.2}$$

$$ITE(x) = 0.4S(x) + 0.6I(x) \tag{3.3}$$

where x is the 16-bit intensity value that is converted into 8-bits.

The pseudo-code implementation for the two stage smoothing operator and ITE transforms as used above are presented in Algorithms 3.1-3.2.

```
1 Standard Windowed Median
2 Input: I[y,x] - Raw image, referenced by y/x; WinSz - Sample
      window size centered at pixel
3 finalOutput: Output image
4
5 Algorithm
6 Function MedianFilter(I,WinSz)
7   For every x and y element in I
8     O= Median(x-WinSz,y-WinSz,x+WinSz,y+WinSz)
9   End For
10 End Function (O)
```

Algorithm 3.1: Generic Standard Windowed Median

```
1 ITE Curve Generation
2 Input: I[y,x] - ITE filtered image referenced by y/x; WinSz -
      Filter window size centered at pixel; SubSample - Filter
      window sampling element
3 finalOutput: O - Output profile
4
5 Algorithm
6 Function MedianFilter(I,WinSz)
7   For every x and y element in I
8     For every SubSample  x and y element in I
9       Cy5sample= Median(x-WinSz,y-WinSz x+WinSz,y+WinSz)
10       Cy3sample= Median(x-WinSz,y-WinSz x+WinSz,y+WinSz)
11     End For
12     Cy5sampleDif=I(x,y)-Cy5sample(x,y)
13     Cy3sampleDif=I(x,y)-Cy3sample(x,y)
14     If Cy5sampleDif <0, Cy5sampleDif=0;
15     If Cy3sampleDif <0, Cy3sampleDif=0;
16     pixlist=add current Cy5sampleDif and Cy3sampleDif
17     sampleMax=maximum value in pixlist
18     O=((sqrt(pixlist))*0.4) +(((1-   1/(pixlist/256+1))*256)
        *0.6)
```

```
19    End for
20  End Function (0)
```
Algorithm 3.2: Subsample Windowed Median and ITE Curve

Algorithm 3.1 is representative of a standard moving-average windowed process. Such a computation averages a given pixel based on the pixels immediate neighbors. As stated, this has the effect of keeping the smoothed pixels similar to their immediate area. Algorithm 3.2 implements the ITE curve functionality as presented in Equation (3.3). Essentially, after applying the 2nd median operator of Figure 3.6 to the image surface, Equation (3.3) remaps the given pixel to the new value (as can be seen in Figure 3.7).

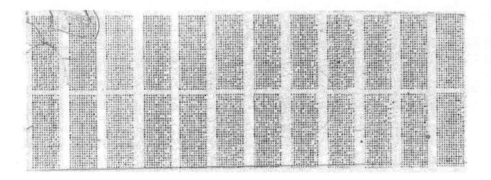

FIGURE 3.7

Blended linear and ITE images showing transform matching

Figure 3.7 presents the same microarray image as used above with the before and after ITE process results blended into one image surface for easier tracking of feature matches. The advantages of this ITE process are particularly clear in this instance. The white regions around the gene blocks have partial background noise present due to the higher inverse transform weighting (on closer inspection it becomes apparent that this "noise" contains valid gene spot pixels). These partial background regions should not pose a particularly big issue for later stages as it is primarily not consistent. Also, the artifact of the upper left region has been reduced to a halo type feature as highlighted. This reduction in the artifact has thus improved the chances of detecting the gene spots that exist in these regions of the image. There are still gene spots in the surface that have been lost to the rescaling process, but critically, their numbers are few and far between.

3.3 Evaluation

In order to test the quality and performance characteristics of the proposed ITE solution, three microarray images were chosen from the 122 available that represented typical analysis problem areas (for ease of discussion, the presented image surfaces are the linear merged channels rather than the raw two-channel surfaces). The top image of Figure 3.8 represents a very "clean," well hybridized experiment result. As can be seen from the image itself, the background element of the surface is substantially lower in intensity than the gene spots. Also in this case, there are no artifacts of any significant intensity present on the surface. The transform should retain most if not all of the gene spots in this image.

The middle image of Figure 3.8 is more representative of a typical microarray experiment. Not only is there significant noise in the image surface, but there are also two large artifacts present that obscure several gene spots to some extent. Here, the ITE transform should remove the bulk of the background noise present, which will render the gene spots with greater clarity.

The bottom image of Figure 3.8 is an example of what can happen when poor biological protocols are used during the experiment process. There is significant artifact noise throughout the surface which is caused by poor hybridization taking place and slide preparation inconsistencies. Applying the transform to this image should have the effect of reducing the strength of the artifacts present while at the same time strengthening the weaker gene spot signals, however there is no guarantee that such a process will result in an improved result. As can be seen from the surface variability in the Figure 3.8, it is difficult to formulate a generalized idea of image classification based on surface quality. As such, images should be treated as individual surfaces which means no unnecessary assumptions are made. Potentially this can generate more work than necessary as the analysis of the bottom image for example will be more involved than that of the top image. However, due to the cost involved per slide as well as the high variability in the surface intensities, such extra work is required.

3.3.1 Experiment results

This section presents the results of applying the ITE transformation technique to the three images of Figure 3.8. The experiments and their significance are explained along with a summary of any improvements brought about by the use of the ITE technique. Rather than show the full size image region in this discussion as in Figure 3.8, a selection of sub-regions will be presented. Each of these sub-regions will focus on a different aspect of the ITE transform functions ability, in this way, it is hoped that a greater understanding of the ITE process will be gained.

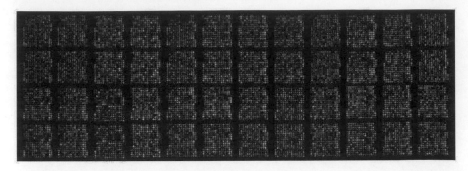

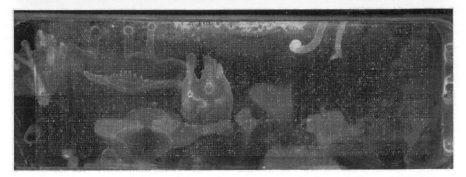

FIGURE 3.8
Representative experiment images with differing levels of quality throughout.
Top image: the good; Middle image: the bad; Bottom image: the ugly

Importantly, note that it is difficult to quantify what is a "good" or "bad" result as such for a given image process unless some form of comparison can be made. Clearly a direct comparison of one process algorithm with respect to another is highly desirable for such a stage, but difficult to acquire in this situation. One such comparison process for example, could compare a human prepared microarray image to that of this ITE result. This is however unfeasible, as the two processes although similar in desired goal, are not actually equivalent (the human can change brightness and contrast values when working). Therefore this process makes subjective comparisons for various regions of given image surfaces. These regions where chosen specifically as they allow the highlighting and examination of challenges seen across the entire test image set. Note that a quantitative analysis will be discussed in Chapter 6 where it makes more sense to do so as the component parts of the CMA framework are used at that stage to generate the final log_2 ratio data (which can be directly compared to their equivalent GenePix generated results).

3.3.1.1 The good

As stated at the start of this section, the good image represents a biologically clean image. The gene spots are well defined and appropriately separated from each other within the meta-block structure. The meta-blocks themselves do not suffer from any overlapping of dimensions or excessive rotation issues. The transformation process should clarify the problem areas in the surface, but critically, these areas will be difficult to appreciate at this point due to the "cleanness" consistency in the individual gene spot intensities. As such, the good image should be considered a baseline or reference image only; therefore it represents a proof of concept for real-world microarray image data. The left image of Figure 3.9 highlights a section of the top image in Figure 3.8, while the right image in Figure 3.9 shows the ITE transform result for said region.

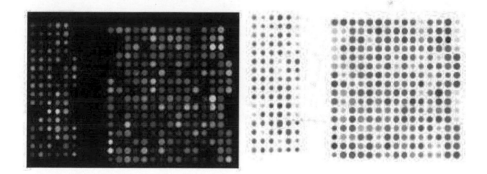

FIGURE 3.9

Comparison of "good" image region before and after applying ITE transform.
Left: Original image region; Right: ITE transformed image region

The dark regions of the left image in Figure 3.9 have very little noise artifact present, also, there are dark regions surrounded by gene spots which one could imagine would infact have gene spots present. From the resultant ITE transformation on the right image, these dark region pixels do contain valid gene spot regions as above. Looking closely in the middle white region of the background of the right image there are slight imprints of background noise from the original surface. Further investigation of this transform process could be made to enhance these weak gene spots to a greater level. However, care would have to be taken with such an amendment process as without the use of some form of texture analysis and gene spot location information such a modification could easily enhance background noise elements (as shall be discussed later). Importantly, the left image in Figure 3.9 is representative of extremely "good" or high quality microarray images and as such, it is expected that the gene spot separations in these cases will be clear. As will be seen in the bad and ugly sections, the benefits of this ITE transform process are more meaningful on images representative of a more typical laboratory experiment result.

3.3.1.2   The bad

When examining image regions more representative of a typical microarray experiment, it becomes quite obvious that, due to the nature of the biological processes involved with the microarray construction process, images with the quality of The Bad appear regularly. The human intervention element of traditional analysis processes have been accepted in the past primarily due to these types of surface anomalies (see Section 2.3 for details). The true task of the ITE transform is to take these various imperfections in an image and render them as insignificant as possible while leaving the gene spot intensities as they are. As in Figure 3.9, the left and right surfaces of Figure 3.10 represent the original and ITE transformed surface regions for the middle image in Figure 3.8 respectively.

There are several regions of interest in the bad imagery. The first, as presented in the left image of Figure 3.10, details the original surface variances as they pertain to the large artifact. It can be seen here that several of the gene spots in this region are either obscured by the artifact fully or have been partially intersected by it. There are also one or two gene spots in this artifact region that have a very similar intensity value to the outlying artifact region. This raises the interesting question as to how well the separation between the foreground and background elements is accomplished.

On examining the right image of Figure 3.10, one can see quite clearly that for the general artifact region in the image, the transform has performed very well. Due to the lack of gene spots between the two partial blocks of genes, the internal region of the artifact has had its intensity values substantially reduced within this background region. Unfortunately, these border regions will also contain intensity values associated with gene spots in other regions

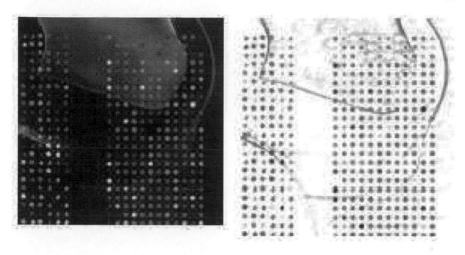

FIGURE 3.10

Comparison of "artifact region" before and after applying ITE transform. Left: original image region; Right: ITE transformed image region

of the image and as such are kept. In regions where the gene spots do exist the artifact has still been reduced, but there are now artifact partials surrounding the genes. In the cases where the artifact boundary has intersected gene spot regions, again the separation is quite affective as the artifact edge has been reduced.

Figure 3.11 presents a close up of the artifact region from the middle image in Figure 3.8 with the appropriate gene spots. The left and right surfaces in Figure 3.11 highlight the original and ITE transform result appropriately, while the middle image presents the complementary pixel movement surface. These movement surfaces are used here to help clarify specific problem areas in the underlying images.

Generally, the gene spots of the left image have similar levels of intensity as the local artifact or background domains, yet the ITE transform has separated the intensities well as seen in the right image. First, looking at the right most columns of the left image, it can be seen that several weak gene spots (the yellow spotted ones,) in this area have been separated cleanly and accurately (in the right image) from their local background. A potential problem for later stages of the proposed framework however, could be in the results of the upper middle region of the ITE transform process. Due to the distribution of the intensities in this local area and the formulation of the ITE response curve, artifact intensities have taken priority over the gene spot intensities. On closer examination of these problem gene spots, it becomes apparent that the intensities of the gene spots are more consistent with background regions rather than the gene spots themselves in these cases.

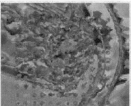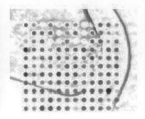

FIGURE 3.11
Comparison of "artifact" image region before and after applying ITE transform. Left: original image region; Middle: movement image region; Right: ITE transformed image region

An understanding of this situation becomes more apparent when one examines the equivalent pixel movement surface for the area as presented in the middle image. Initially, one is struck by the consistent strong channel biases that can be seen in this region. Then, looking closely at the artifact area in the middle image, it can be seen that coupled with some strong channel biases, the individual gene spot surfaces are littered with strong highly variable pixel intensities. This high surface variability means that some gene spots in the image will consist of intensities more similar to background. These extremely noisy surface regions are typically rendered as "white" space within a gene spot region in the right image.

If these "white" region intensities are examined more closely it is difficult to separate them from their local background (indeed, such gene spots are usually marked as corrupt by the operator of the GenePix process). If a region with a local background of different intensity levels to that of the images of Figures 3.10-3.11 is examined, and if this new regions intensities are reversed (gene spot intensities from Figure 3.11 are similar to background intensities of this new region), this intensity distribution effect can be readily understood.

In Figure 3.12, the left image shows an original microarray region of middle image of Figure 3.8. The middle and right images of Figure 3.12 present the pixel movement and ITE transform result surfaces respectively.

Examining the transform result of the right image in Figure 3.12, we see that the spotted gene spots have been enhanced as expected, considering their similarity to the local background regions. The yellow spotted gene spots on the other hand have been poorly identified. Such poor gene spot identification is essentially caused for two reasons:

1) The weak intensity gene spots in this case have an intensity that falls below that of the local background, meaning in essence, the gene spots are negatively expressed. Such negative regions are lost track of during

FIGURE 3.12
Comparison of "high background" image region before and after applying ITE transform. Left: original image region; Middle: movement image region; Right: ITE transformed image region

the ITE process as the background region itself is by definition greater in intensity.

2) The classification of these weak intensities is made more complicated as they are similar to background pixels in other regions of the surface. This reversal of fortune means, in a static curve system, that the weak intensities are classified as background in these regions.

This intensity distribution process can be seen more readily in the middle image of Figure 3.12. Note how the problem gene spots of the right image are essentially surrounded by mid-to-high intensities in the right image (and as such, their definition is lost). This suggests that the ITE response curve should be dynamic in nature rather than static as it currently is.

3.3.1.3 The ugly

As seen in the right image of Figure 3.8, this ugly microarray experiment suffers from several intense artifact problems not seen in the middle and right images of Figure 3.12. It becomes apparent from the intensity distributions in the image that the artifacts have a significantly higher intensity value than in the left and middle images of Figure 3.12. There will be significant numbers of gene spots that are hidden from view due to such intensity mismatches. More importantly, there will likely be gene spots lost from the experiment as a result of the biological binding issues the researchers faced. This original image will also contain all of the issues as previously mentioned in relation to the left and middle images of Figure 3.8. The left and right surfaces in Figure 3.13 present the original and ITE transform result appropriately, while the middle image presents the complementary pixel movement surface.

However, with these issues in mind, from the transform result of the right image in Figure 3.13, the ITE process does not seem to have suffered too badly from any detrimental effects of these consequences. Indeed, excluding the missing gene spots as mentioned above, only the outer edge of the artifacts have been pulled into the resultant for the most part, this means the internal

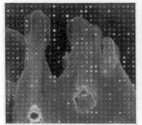  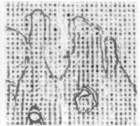

FIGURE 3.13

Comparison of "extreme artifact" image region before and after applying ITE transform. Left: original image region; Middle: movement image region; Right: ITE transformed image region

regions of the artifacts have still had their "noise" substantially reduced. This reduction has resulted in the enhancement of several of the weaker gene spots; this is most notable in the "eye" region of the artifact. Examining the pixel movement (middle image of Figure 3.13), one can see that the artifact region is indeed littered with small areas of consistent intensity. It can also be seen from this imagery set that the artifacts edges are preserved with such clarify as their surrounding pixels have consistent yet different intensity spread. Such prevalent artifact partials could create negative consequences for downstream processes. One way to subdue these partial edges would be to apply a spot template across the entire surface area, with negative agreement pixels set to a zero value. This could however render those gene spots covered by the edge artifact as nonexistent. A better idea therefore would seem to be the softening or dampening of this edge signal. In this case, gene spots close to the edge in question would also have their intensity reduced, but this would still be a more favorable result.

As hinted at in the results of Figure 3.13, it is the hard edges (as well as the response curve formulation) of the surface that causes these outer edge artifacts to remain in the resultant transform image. It could be argued that this feature helps with the retention of the gene spots themselves, as they are essentially small transition areas. Telescoping or intensity stretching techniques could reduce the edge information retained, but should be used with caution. With respect to this edge issue, Figure 3.14 presents an area of the surface that has multiple soft edge artifacts, where the left image highlights a region of surface that has three distinct dye bias artifacts, and the right image presents the ITE transform result for the region.

The top left and central artifacts in the left image of Figure 3.14 are biased to the red channel data (as is the image generally) whereas the lower left artifact is biased to the green channel. Two points become apparent with analysis result of the right image. First, this channel biasing does not have an

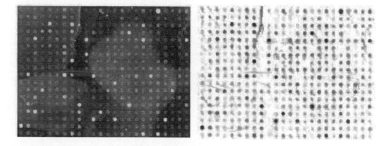

FIGURE 3.14
Comparison of "bias" image region before and after applying ITE transform.
Left: original image region; Right: ITE transformed image region

obvious direct impact on the ITE process (excluding the hard edge aspects as mentioned). The second point re-emphasizes the fact that the soft edge transitions are handled more gracefully than their hard edge cousins. This would suggest that along with the dynamic curve aspects of the ITE improvement process, a gene spot morphology characterizer should be created to help with the sorting of valid edge detection in situations where there is significant noise present.

3.3.2 Strengths and weaknesses

The ITE transform process, as highlighted in Figure 3.6, initially executes two median average operators over the images surface. These operators have the positive effect of removing small inconsistent values from the image. For example, a fine scratch across an image's surface will be substantially reduced in intensity with the application of such a process. However, there is a negative aspect associated with these operators. Essentially the median works by setting the current pixel of interest to an average value of the surrounding region. Problems can arise when the surrounding region is not consistent in surface variance with the image as a whole.

Another potential problem here is the way in which the second level median works. Pixels in this region window are sampled at every fourth entry. This means generally, the background element is sampled slightly more than the foreground. This sampling can have the effect of reducing the internal gene spot region's intensities which could have negative effects. For example, if the surrounding regions of the pixel of interest do not represent a similar intensity range, the gene spot itself could blend into the background. The positive side of this aspect however can be seen in Figure 3.7 with the reduction of the artifacts internal regions.

Once these two median operators have been applied to the two image channels (in this case), the maximum pixel in a given region is then chosen as the representative of that pixel across both channels. The ITE transform is then applied to this representative pixel. This maximum operator is applied to help with the reduction of memory and storage requirements. It would be more advantageous to execute the ITE functionality directly onto the second level median results for the number of channels as required. This would build an image that is essentially more accurate. The processed ITE results could then be combined in some way to retain the reduction benefits of the current implementation.

Overall, the current implementation of the ITE transform works very well over the available microarray images. However, as suggested by the noise elements present in the imagery of Section 3.3 the response curve does exhibit some weaknesses. The current curve is static in nature and was derived from the results of an intensive study of the images. As seen in the subsequent figures, there are several intensity regions within a typical microarray image that can be representative of a gene spot. Clearly, one improvement that could be made to the ITE process is the creation of a dynamic response curve rather than the current static one. Dynamic is used in this context to mean the curve responds in some way to the underlying surface.

The gene spot intensities of a typical microarray image can roam over a large range of the possible 65535 values. It is conceivable that in some areas of the image a gene spot could have an intensity range associated with the background in another region (as seen previously). The response curve could be generated from a selection of curves (rather than a hybrid of two) that highlight multiple frequencies associated with the gene spots. Alternatively, a texture analysis type process could be implemented that examined the regions as defined by the two successive "filter windows." This window's shape could be synchronized with a particular *region of interest* (ROI). This window is initially set to 5×5 pixels to remove local outliers and then to 57×57 pixels to generate a background estimate. Changing these dimensions to fit a problem area more closely will yield more clarity in the area. By way of example, suppose that the PCC component loses a large area of signal due to major surface variance. PCC could request that the ITE generate a new view of this ROI which would clarify the area better than that generated by the static pass process. This would have the knock-on effect of more accurately identifying low level type features as the range of probable intensities within the area will have shifted to a local norm.

Another approach would be to harness the surface data from the two~four channel images individually. These channels exhibit very high similarity (as in fact they are the same surface due to the scanning process using two slightly different scan frequencies) and as such there are essentially two surface measures for any given pixel. This channel information could be used to generate a warp image defining the individual pixel confidences or affiliations within a given gene spot boundary.

3.4 Summary

This chapter has examined the effects of applying image transformation techniques to microarray image data. It has shown that when the correct filtering approach is applied to the imagery, the gene spot elements can be greatly enhanced. This enhancement leads to a clearer understanding of the variability at play within an image's surface. This understanding in turn allows the rendering of the effects of background variability and artifacts on the gene spot expressions themselves to be reduced in severity.

Although the three example images presented in Section 3.3 demonstrated the *Image Transformation Engine* (ITE) method produces good results, there is still room for improvement. One area of improvement could be the harnessing of more image information by processing the images channels separately as mentioned in Section 3.3. The two channels in this case share many similarities and it would make sense to try and utilize this data to a greater degree. Alternatively this future work could pursue the idea of changing the static nature of the response curve. Currently, the curve exists as a hybrid of the inverse and square root functions. Although this hybrid process has performed very well over the available image data, problems were seen to crop up in situations where the gene spot intensities were very close in value to background noise. A dynamic implementation of the ITE curve could be created such that greater use of the movement information of the pixels are utilized.

An alternative to a dynamic response curve implementation would be to tie a *particle swarm* (PS) type system into the ITE process. Such a PS process would essentially be filtering the image surface according to a consensus of agents. Rather than processing median or mean operators for the image area, these agents would essentially act as these functions. Thus a dynamic surface map would be rendered rather than the segmented map at present.

An enhancement process has been created such that the raw microarray image surface can be cleaned up somewhat and the probable gene spot regions appropriately enhanced. The next chapter focuses on a technique that approaches the enhancement of gene spot regions from a different direction. This different approach can be thought of as a "belt and braces" affair in that the framework essentially gets two bites of the problem space. Whereas the ITE process assumes the gene spots reside in a certain intensity range, the novel *Pyramidic Contextual Clustering* (PCC) technique presented in the next chapter assumes that gene spot pixel's are close together. In essence, the

PCC approach uses a given pixels neighborhood (across the whole surface) to identify the probable gene spots. Similar to the ITE approach sometimes removing gene spots from the surface if they are very similar to the background intensity so too the PCC technique will identify background pixels as belonging to a gene spot occasionally. Both the ITE and PCC methods are powerful in their own right, but when combined, they reinforce the confidence of a particular gene spot region.

4

Structure Extrapolation I

4.1 Introduction

Chapter 3 identified the problems involved when working with full size complementary deoxyribonucleic acid (cDNA) microarray slides. These problems primarily revolved around the high signal variability as generated by the existence of various image artifacts. The chapter detailed the benefits of using a multi-view process on the images to emphasis various aspects of their surfaces. An enhancement of this nature allows for the partial removal of an image's artifacts and thus re-focuses the image into the more likely regions of interest (the gene spots).

The *Pyramidic Contextual Clustering* (PCC) algorithm presented in this chapter extends this enhancement idea to the next level. The PCC process is designed to emphasize the probable gene spot regions via a different approach to that of the *Image Transformation Engine* (ITE) technique of chapter three. Although the gene spot positional information as rendered by PCC is partial in some areas of the newly generated image (as per the ITE), these partial gene spots are still of benefit to the underlying task of gene spot identification. When used in combination, these overlapping ITE and PCC processes should render better image knowledge than the processes would individually.

Current research work in the field of large datasets with clustering approaches relates to the processing of large database structures; examples of which can be seen in the development of algorithms such as CLARANS [151], BIRCH [230] and CLIQUE [3]. Importantly, note that although these individual papers refer to large datasets as containing many gigabytes and terabytes of information, in practice they process only a fraction of this. CLARA [118] is an adaptation of the original partitioning around medoids (PAM) clustering algorithm designed specifically to process larger datasets. CLARANS is a further modification of the CLARA algorithm and is based upon a randomized search process using parameters to limit the number of neighboring points and local minima that are explored. The authors of the BIRCH algorithm highlighted that by limiting their search process as in the CLARANS implementation, BIRCH may not discover a real local minimum. Therefore, the BIRCH authors presented an algorithm designed to minimize the I/O costs while working with limited memory and thus attempted to show that BIRCH is consistently superior to the above methods. The CLIQUE algorithm was designed to answer several special requirements in data mining that the authors felt were not adequately addressed.

Clustering [109,110,133] as related to microarray data is typically focused at generating a clustering of the resultant gene spot intensity values themselves rather than as a means of finding the gene spots per se. A good example of this can be seen with Balasubramaniyan [18] who searched for co-regulated genes. The most common technique employed for such analysis work is that of hierarchical clustering [147,213] techniques. Eisen [70] described a hierarchical clustering algorithm that used a greedy heuristic based on average linkage methods. Leach [128] presented a comparative study of clustering techniques and metrics for gene expression levels. Lashkari [126] described an experiment setup for human T cells analysis, while a Bayesian network approach was proposed by Friedman [90] that described the interactions between the genes in a microarray. Jansen [113] investigated techniques relating gene expression data to protein-protein interactions.

However, with a typical microarray image consisting of over twenty million–plus pixels, current clustering methods simply cannot be directly applied to the image data as a whole. Early research work in this field can be seen in papers [32, 34–36, 149, 227], where the authors looked at the use of clustering as applied to microarray imagery specifically. To overcome computational expense issues associated with clustering the full image, Bozinov and Rahenführer [32] proposed an abstraction of the k-means [139] technique whereby pre-defined centroids were chosen for both foreground and background domains, to which all pixel intensities could be assigned. This is to say that one centroid was chosen to represent a gene spot signal with another centroid chosen to represent the noise or artifact. Unfortunately although traditional k-means is able to choose centroids according to the dataset's characteristics, this approach is inherently biased towards outlying values (saturated pixels for example), and not the true region of interest (the foreground pixels). In Nagarajan [149] the issue of clustering the full slide was not addressed specifically as the authors were interested in the effects of clustering the individual gene spots. Yang [52, 227] presented a general review of this area, detailed other manual methods and proposed a system called SPOT that had some improvements over previous methods.

Other methods present ideas along similar themes, for example, the application of wavelets [209, 210] and Markov random fields [115–117] show great promise, however, at this time they have only been used on what would be classified as "good slides," and even then, not on realistically dimensioned data. If these techniques fail to determine the location of just one spot, the system would fail, thus having to fall back on user intervention in order to recover. Indeed, when processing large scale datasets, Berkhin [24] classified the current solutions into three groups: incremental mining [212], data squashing [64, 65], and reliable sampling [143] methods. The main drawback with these implementations is that, by reducing the number of elements in the datasets, important data will have been lost and so there is a need for developing techniques that can be scaled to these large scale image problem areas directly.

With completion of the Data Services stage of the *Copasetic Microarray Analysis* (CMA) framework, the investigation shifts its focus onto finding possible gene spots within the image. The *Pyramidic Contextual Clustering* (PCC) algorithm examines the full surface area of the microarray image at the pixel level. These surface pixels are compared together (according to some proximity criteria), and assigned into one of the two groups, which represent either signal (foreground) or noise (background). The proximity criterion is scaled up through all iterations to give the algorithm as much "pixel spread" information as possible. Once the proximity criterion is greater than the dimensions of the raw image a consensus result is generated. This consensus image along with the PCC "time slices" are used throughout the CMA framework as the building blocks for acquiring image knowledge.

This chapter aims to explain how the PCC technique of Figure 4.1 is able to process a full microarray image (something not possible with traditional clustering techniques) with the mechanisms involved in breaking up the pixels into their foreground (the gene spot) and background groups.

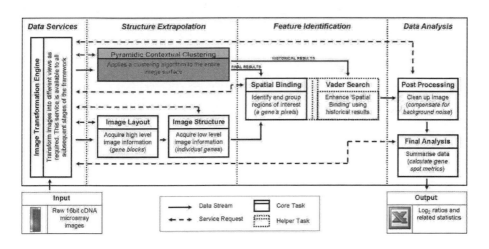

FIGURE 4.1

The CMA framework

The PCC process combines spatial knowledge of the ITE generated image surface with these ITE foreground enhanced intensity distributions, highlighting three important objectives for the study presented in this chapter.

1) The proposed solution should be able to cluster a raw microarray image

directly. Rather than sub-sampling or removing certain pixel elements from the image surface, the process must somehow harness all of the pixel intensity information. Such individual pixel acquisitions will render greater accuracy to the output image of the process.

2) Due to the noisy nature of a typical microarray image (see Section 2.3 and the image of Figure 4.6 for details) the solution should be dynamic to a local pixel area. More critically, the solution should have some concept of time. This means as the clustering process is executed by a sequence of steps; the internal decision processes involved in these individual time steps should be recorded.

3) The proposed solution should execute in a reasonable amount of time and thus be of low computational time complexity. Also, this clustering process must operate without the need for domain specific knowledge from the outset.

Based on the three objectives, Section 4.2 presents a broad overview of the proposed solution to this scaled clustering problem as it pertains to the CMA framework. Section 4.3 evaluates the solution with respect to synthetic and real-world microarray data. Section 4.4 concludes the chapter with a summary of the challenges faced when applying clustering techniques to the analysis of microarray imagery.

4.2   Pyramidic contextual clustering

4.2.1   The algorithm

Initially the PCC component systematically divides up the microarray image into spatially related areas (normally very small grid squares), as is common in the artificial intelligence [177] and clustering [148] communities. Each of these grid square areas is then clustered by using a traditional technique such as k-means [139] or fuzzy c-means [66] with the result stored appropriately. Once this initial clustering process has finished, representatives are calculated for each of the clusters that exist, with these representatives then further clustered in the next generation. Such a process is repeated until all of the sub-clustered groups have been merged into one of n groups. The main concept of this PCC technique is demonstrated with a simple application of the method as described above, in Figure 4.2.

In the input layer of Figure 4.2, it can be seen that there are twelve objects which are to be clustered (the different shapes represent values rather than pixels per se). With traditional clustering techniques, these twelve objects would be treated as one set (rather than the three sets in the figure). This

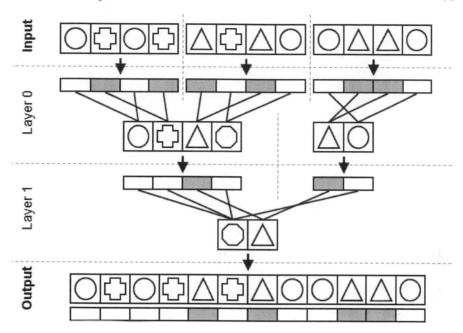

FIGURE 4.2
PCC process diagram

set's objects would all be compared together with a likely output being one group of triangles and one group of a mixture of circles and crosses as seen in the output layer of the figure.

The PCC technique would produce a final output that is very similar to this traditional output. However, the PCC internal functionality takes a different approach in its analysis. PCC divides this set of twelve objects into sub-groups of a given size (into three sets of four as shown in the input layer of the figure). Each of these sub-groups would be clustered into one of two classes (represented by the checkerboard pattern or lack thereof, within each layer below the shapes). In layer 1, the representatives (as generated from the raw data in layer 0) for each of these groups have been clustered together. The final result can now be calculated by re-traversing the pyramid structure and tracking the shaded members.

Studying the three groups in this example closely, it can be seen that the shapes (which represent differing intensity values in this case) are chosen to illustrate how this sub-clustering can provide useful contextual information. The first group of circles and crosses can easily be clustered as two separate shape classes with no problems (as shown in layer 0). With the introduction of triangles in the second sub-group, it can be seen that the circle and cross are now clustered together; here this is the optimum arrangement due to shape similarity. By layer 1, the groups are the same as a traditional clustering algorithm, with all circles and crosses in one group and the triangles in the other. Unlike a traditional clustering algorithm however, information from the previous layers can now be used to ascertain the fact that in some situations certain objects should not have been grouped together. Here the circles and crosses could have formed two distinctive groups if it were not for the presence of the triangles. Importantly, note that this information is context specific, and relies on a spatial dataset, such as an order dependent vector (an image).

The pseudo-code implementation of the PCC technique as described above is presented in Algorithm 4.1. Note that this presentation has been based around clustering an image (when utilizing traditional techniques) into two groups containing either signal or artifact objects for reasons of clarity.

```
 1  Inputs
 2  Image: The n bit tiff image
 3  WindowSz: The size of the window to use when sub-dividing each
        level
 4  ClusterMetric: traditional clustering metric (k-means, fuzzy c
        -means, ...)
 5
 6  Outputs
 7  Signal, Noise: Mutually exclusive co-ordinate lists for every
        pixel in Image
 8
 9  Algorithm
10  Function PyramidicContextualCluster(Image,WindowSz):Signal,
        Noise)
11  Width=Image.width
12  Height=Image.height
13  FirstPass=true
14  While (Width > 1 AND Height > 1) do
15      NewWidth=Width/WindowSz
16      If (NewWidth < 1) Then NewWidth = 1
17          NewHeight=Height/WindowSz
18      If (NewHeight < 1) Then NewHeight = 1
19          Create Signal[NewWidth][NewHeight] as empty Co-ordinate
                list
20          Create Noise[NewWidth][NewHeight] as empty Co-ordinate
                list
21      For (y = 1 to NewHeight)
22          For (x = 1 to NewWidth)
23              If (FirstPass = TRUE) then
24                  Resultant=ClusterMetric(Image,x,y,x*Window,y*
                        Window)
25                  FirstPass=false
26              Else
27                  Resultant=ClusterMetric(Signal,Noise,x,y,x*Window,
                        y*Window)
28              End If
29              Signal[x][y]=Resultant.Signal
30              Noise[x][y]=Resultant.Noise
31              Width=NewWidth
32              Height=NewHeight
```

```
33        End For
34      End For
35 End While
36 End Function(Signal,Noise)
```
Algorithm 4.1: PCC Pseudo-Code

Note that the use of the two binary image surfaces *Signal* and *Noise* have been presented to store the allegiance of an image pixel to a specific cluster such that understanding of the PCC concept is aided. The implementation can easily be modified to work with multiple groups. Importantly, note that in the implementation, the ClusterMetric function (k-means for example) is executed three times in order to take the best result based on intra-cluster similarity scores due to the stochastic nature of the algorithms. This "best result" is also assisted by the clustering event starting from a 2^2 sample *WindowSz*, in order to reduce initiation biases.

4.2.2   Analysis

As stated, one of the fundamental strengths of the PCC process over traditional clustering techniques and that as proposed by Murty and Krishna [148] is PCC affectively renders transparent the ClusterMetric's internal decision process. This is achieved by way of the pyramidal layering concept (as seen in Figure 4.2) which gives the CMA framework the ability to review historical (or time-slice) image information. These "time-slices" further enhance the improvement of the final microarray image clustering output. Traditional clustering algorithms lack this historic information capability and the problems that arise from this can be illustrated with a simple test image. It may seem counter intuitive to propose a clustering process on such large data; however, such a step is advantageous in differentiating between signal and artifact features of the raw surface. As mentioned previously, a possible improvement to this stage would be to incorporate a multi-resolution process into the overall component. The advantages of the historical information from PCC and the storage/efficiency improvements of the multi-resolution approach could be harnessed more appropriately.

The test image has both flat (ideal) and gradient (realistic) based background surfaces and these are presented in Figures 4.3-4.4. The test image has dimensions of 100×100 pixels (such that traditional processes can be compared to PCC directly) with well defined signal (the arrows) spread throughout the surface. All of the pixels in the image that do not represent signal can be classified as noise. These noise pixels consist of very low intensity random values such that the separation of the two pixel classes is a trivial task. The noise pixels of the realistic image surface on the other hand consists of the same low intensity random values with an increasing ramp function applied over them. Such a ramp generates a smooth gradient signature across the height of the image surface such that the task of signal and noise separation becomes non-

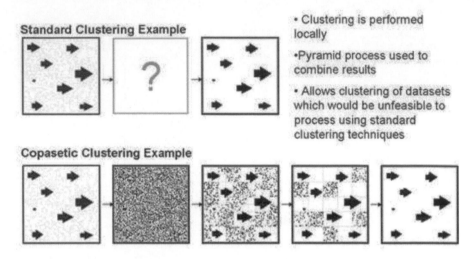

FIGURE 4.3

Comparison of traditional and PCC processes: clustering for ideal data

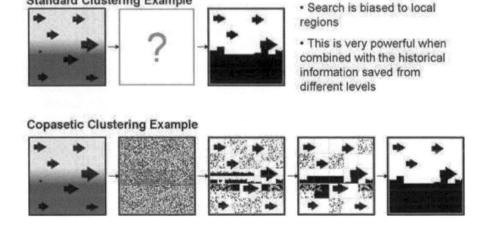

FIGURE 4.4

Comparison of traditional and PCC processes: clustering for realistic data

trivial as the image surface is parsed. The question marks in the two central boxes of Figures 4.3 and 4.4 denote that traditional clustering techniques are blind to their internal decision processes.

In the case of the ideal background surface, Figure 4.3, the blindness of traditional techniques has no direct consequence to the accuracy of the produced result. The lower plots of this image depict the PCC process flow for the same input image. It is not surprising that the PCC output is very similar to that of the traditional technique (as it should be), but critically, the ClusterMetric's internal decision processes have now been laid bare.

This critical aspect becomes clear when focus is shifted to a more realistic test image. The plots of Figure 4.4 represent this realistic microarray region. This figure has a background surface that tends to a darker value from the middle of the image space. In a typical microarray image, characteristics of this nature can be seen across the entire surface. Due to this gradient intensity shift, noise pixels in one part of the image surface could well represent signal pixels in another. The traditional clustering approach has failed to ascertain the signal pixels located in the lower (darker) regions of the surface as a result of this gradient artifact. The final traditional output has, as a result of such artifacts, clumped the entire lower region of the surface into the signal category. In essence, traditional clustering states there are four arrows in the image surface, with one of them being substantially bigger than the rest. Clearly, checks could be made to quantify such a result but how, exactly, can this situation be rectified?

If attention is focused to the PCC approach for the realistic image (the lower plot), it is seen the final output is similar. However, the real advantages of the PCC approach over that of traditional techniques can be seen in the middle plots of the lower images. These internal image surfaces (2⌣4) clearly show that the PCC process has correctly (in all cases) identified or separated the signal from the background class. This is evident even in the case of the extremely small signal element located in the middle left of the image. Clearly the "time-slice" concept of the PCC approach not only adds significant benefit to large scale cluster image analysis, but also any task involving traditional clustering methods.

There are two ways in which these PCC results can be utilized. First, simply take the final output result; which is equivalent to the result from traditional clustering techniques. Second, (and of more benefit) is to produce an amalgamated result from a consensus of the "time-slice" layers as shown in Figure 4.5. This consensus result would hold more pixel details from the original input surface than would have otherwise been possible using traditional methods.

The consensus layer can be generated by calculating the mean value for each pixel through the previous "time-slice" layers and then discredited using a threshold of 0.66 (agreement between two thirds of the slices). For a dataset which consists of two or more clusters a median could be used to the same effect, although the use of more complex methods of forming an agreement are not ruled out.

The final output from this PCC process then, is an amalgamated consensus layer as shown in Figures 4.6-4.8, which give the microarray surface after the PCC transform has been applied. This consensus layer represents a good compromise between the original raw image features and a gene spot (feature) only image. Figure 4.6 is one of the original microarray images as seen throughout

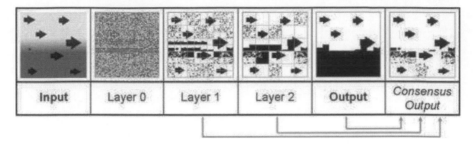

FIGURE 4.5
PCC consensus surface generation process

the book so far, highlighting the various artifact and gradient background features as mentioned. Images in Figure 4.7 are the resultant "time-slice" layers as produced by the PCC process for this specific image surface. The consensus image surface (as computed by means mentioned earlier) is shown in Figure 4.8, with an appropriately enlarged region highlighted.

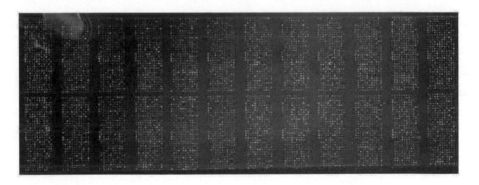

FIGURE 4.6
Original microarray image

From the discussion of the PCC process and emphasized in the figure above, the results of the PCC process are not perfect. Similar to the ITE process, various partial and missing gene spots have been generated and lost respectively throughout the images surface. Nevertheless, the resultant PCC output represents a significant improvement for the task of finding gene spots from the surface than that of the original image surface seen in Figure 4.6.

Note that one could argue that this PCC concept is actually part of the ITE process as discussed in Chapter 3. However, as originally highlighted in Section 2.5, the CMA framework's components were designed such that an

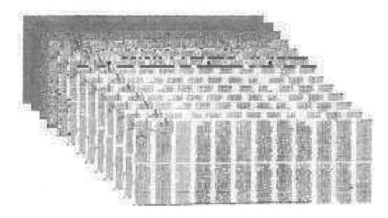

FIGURE 4.7
PCC "time-slice" layers

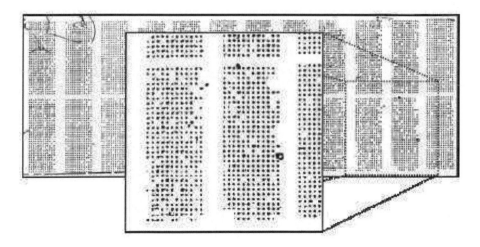

FIGURE 4.8
PCC consensus layer

overlap exists between them. This overlapping functionality helps to enhance the framework's effectiveness as essentially it means that slight errors propagated within a component can be accounted for to some extent by later stages. Indeed, as shall be shown in Chapter 6, the results of PCC and those of the ITE process are used together to generate an accurate picture of the individual gene spots in a full microarray image.

4.3   Evaluation

4.3.1   Search grid analysis

The WindowSz parameter of Algorithm 4.1 controls how the input image will be divided into spatially related subgroups and has an effect both on the overall algorithms performance and the quality of the final results. This parameter can take any integer number in the range [2, dims], where dims represents the dimensions of the image itself. As a general rule the smaller the window size the higher the granularity of the result, however this comes with a penalty in speed. As the window size is increased the speed of the algorithm will improve. As this window approaches the size of the image set the algorithm's performance will converge to that of the underlying clustering technique, until the window encompasses the whole image at which point it functions as the underlying algorithm.

Another reason to keep the window size small is when processing spatially orientated data. If the window size is larger than the artifacts present in the image (as seen in Figures 4.6-4.8) the benefits of processing the spatial information will be significantly reduced. This extra information can help when constructing a consensus result such as the agglomerative function presented previously in this section. This WindowSz parameter then is defined in

$$\text{WindowSz} = \lceil \sqrt{2c} \rceil \tag{4.1}$$

where c is the number of clusters used.

This is provided for guidance only as theoretically the window size only needs to contain $c + 1$ objects. From a performance perspective the previous formula generates a square window with at least $2c$ objects. It should be noted that clustering algorithms which can have clusters with zero elements are not governed by the same constraints. Figure 4.9 presents an analysis process that specifically examined the effects this WindowSz parameter would have on the results of the PCC output. Essentially, the WindowSz parameter was increased in size while clustering a test image. In this case, the test image has different shapes embedded into the surface as the study was particularly interested in the effect of splitting the WindowSz parameter over a shape region. Therefore, the morphology of the shapes involved had to be such that subtle effects could also be noticed.

As can be seen in the figure, increasing this WindowSz parameter value has the effect of reducing the number of layers generated by the PCC process (and in turn the time required completing the clustering task). Examining the row's internal images (images between the Layer 0 and final row Layer respectively) gives insight into the justification of this theoretical limit as the finer quality of the internal layers regions becomes reduced as the WindowSz

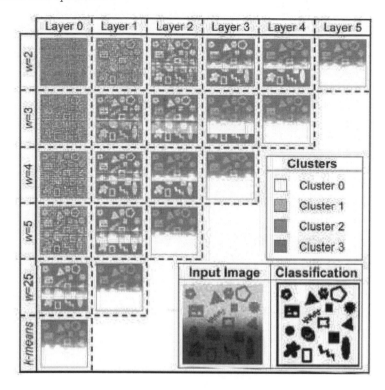

FIGURE 4.9
PCC process with increasing WindowSz value

value is increased. Also, as the parameter is increased the PCC process tends
to produce a final output more similar to a traditional clustering approach.

4.3.2 Synthetic data

In order to give some idea of the performance characteristics of the PCC algorithm, the system is fed a selection of ideal (flat) and realistic (gradient)
synthetic data prior to the real microarray image data. Figures 4.10-4.11 provide *the synthetic and realistic microarray images.* The various shapes in the
surface images of 4.10 signify the patterns or signals of interest. The top row
in Figure 4.10 (uniform) illustrates a selection of images (with dimensions
500×500 pixels) with various degrees of solid white noise. In the uniform set
of images the objects have been rendered with an intensity value of 50 percent
while the background was originally set to 0 percent noise. Over the course
of the series the background noise steadily increased in increments of 5 percent. Such uniform data represents the baseline dataset, that is, it allows the

comparison between the different techniques on what could be classified as like-for-like.

The gradient surfaces (bottom row in Figure 4.10) have had a gradient filter applied to them in order to render a more realistic microarray characteristic surface area. Over the course of this dataset the background noise surface was steadily increased in increments of 5 percent. This gradient filter has the effect of making the lower-region background pixels more similar to the intensity of the shapes than those of the upper regions. The filtering represents a "realistic biological" surface at a much reduced scale. These datasets represent a good middle ground in preparation for examination of true microarray image data as seen in Figure 4.11. Figure 4.11 is a grey composite of a typical microarray surface, which highlights the potential intensity differentials throughout the area. Note particularly that the surface variance of the image is such that background regions in the black triangle are similar to gene spot regions in the white triangle.

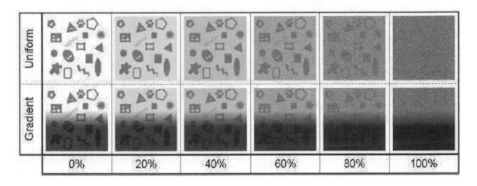

FIGURE 4.10
Synthetic datasets applied during PCC testing

As we know the accepted morphology for the synthetic imagery involved (a perfect mask of shape pixel positions has been calculated), the PCC technique can be directly compared to that of the traditional process by calculating the percentage error between the two approaches. This percentage error is defined in

$$\text{MSE} = \frac{1}{mn} \sum_{i}^{m} \sum_{j}^{n} \|I(i,j) - M(i,j)\|^2 \tag{4.2}$$

where m and n are the image dimensions and I and M are the actual input and mask images respectively.

Figures 4.12-4.15 present the various performance characteristics of traditional and PCC based clustering as executed on the synthetic imagery, i.e.,

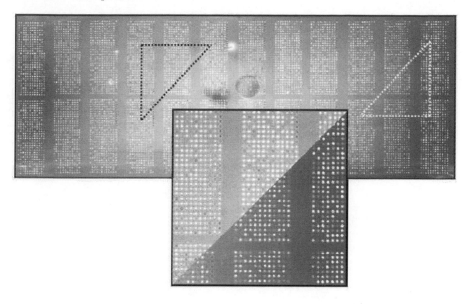

FIGURE 4.11

Representation of microarray surface variances

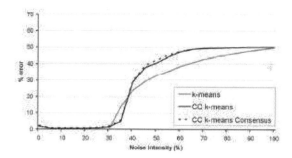

FIGURE 4.12

Uniform k-means

results of k-means and fuzzy c-means clustering on synthetic image set (Figure 4.10). Each graph in the figure consists of three plots: the grey line shows the traditional clustering technique when applied to each image, the black line shows the performance obtained when the PCC technique is applied and finally, the dotted line shows the results when a consensus of the historical PCC information is utilized. Figures 4.12 and 4.14 present the results as achieved by the k-means algorithm, while Figures 4.13 and 4.15 show the fuzzy c-means result over the uniform and gradient image data respectively.

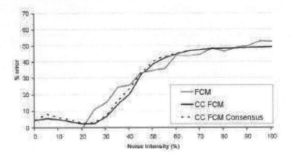

FIGURE 4.13

Uniform fuzzy *c*-means

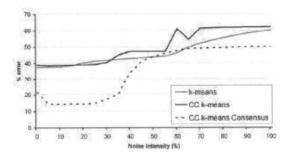

FIGURE 4.14

Gradient *k*-means

Figures 4.12 and 4.13 present the percentage error when the two clustering techniques are applied to the uniform based synthetic datasets. As this dataset contains a uniform spread of noise elements it was envisioned that the results of the experiment would be similar between the techniques. Due to this uniformity of the background, foreground pixels are clearly defined and hence easier to cluster. This clarity of separation is clearly seen by interpreting the dotted line in the figures. Essentially, the historic ability (dotted line) of the PCC based clustering technique has failed to draw any more knowledge from the uniform dataset (the solid black and dotted history plots are nearly identical). In the case of uniform synthetic data, PCC does not really have an advantage over traditional approaches (other than scaling the clustering techniques to larger datasets than would have been possible that is).

However in many real datasets such uniformity would not be the case, rather, variations in background (as seen in Figure 4.11) would mean pixels could only be classified correctly when taken in the context of their local region. Currently traditional clustering techniques fail in this situation, due to

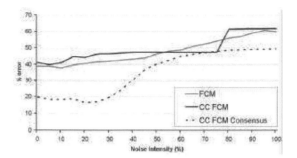

FIGURE 4.15

Gradient fuzzy *c*-means

the fact that every pixel is treated in isolation, inherently PCC is spatially aware and therefore able to capture such contextual information. Figures 4.14 and 4.15 hint at this spatial improvement capability by presenting the percentage error when the techniques are applied to the more biologically realistic gradient based synthetic data. Here, both the traditional and PCC clustering approaches have performed on a par with one another, which is the expected result (as seen by the grey and black lines respectively).

However, when the historical information (dotted line) is brought into the fray, there is substantial improvement across the entire series. Indeed, it can clearly be seen that the historical data reduces the initial noise error generated over the 30~35 percentage noise error images. This noise then starts to rise over the 60~70 percentage noise images quite quickly before leveling off again. Importantly, note that the PCC approach (black line) has not deviated significantly from the traditional approach of the grey line (which one would expect in this case). Note also that the spike in the black line plot (around the "Noise Intensity" 60 percentage mark of the *k*-means image) is caused by a large inverted (signal and noise classifications reversed) search-grid block. To further illustrate the benefits of these historical "time-slice" layers over traditional approaches some example output images are shown in Figure 4.16.

Although there is still noise present in the PCC generated output images of the figure, the objects in the image have been clearly defined and could be used for further analysis such as edge detection. In contrast, the traditional clustering results have lost this information completely and nothing further can be done to clarify the lower region without re-analysis by some other method.

4.3.3 Real-world data

The gradient based image data in Figure 4.11 were designed to be sympathetic to the characteristics of a true microarray image. Although this realistic like

FIGURE 4.16
Example output synthetic images

image data has appropriate high-level structures, there can be no replacement for the real thing. Figure 4.17 represents a typical clustering result as computed over true microarray data by traditional and PCC approaches. The top layer 0 box in the figure is a cropped region of a real microarray image surface. The region is restricted to 500×500 height and width dimensions due to traditional approaches (k-means for example), only being able to cluster such a region size (on available hardware). The top layer 1 box represents the output for the traditional and PCC clustering processes. The bottom row layers are the resultant PCC results for the same input image region.

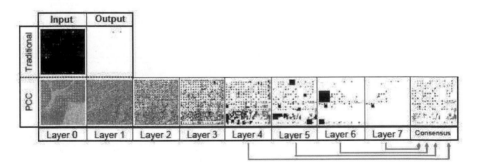

FIGURE 4.17
Real-world PCC results for microarray image region

From the traditional approaches layer 0 output, the strongest gene spots have been identified while the remaining genes are lost to the process. Indeed, there exists a large region of the input image's gene spots that have failed to be clustered satisfactorily. As mentioned, the two strongest gene spots found in the surface region have the strongest intensity of all the gene spots in the image. The not so strong gene spots slowly become masked out by their stronger peers and thus are removed from the final quantification step.

The PCC approach produces a final output that is very similar to that of the traditional methods, however due to the transparency of the technique; this weakness can now be accounted for more satisfactorily. From the bottom layer 0~layer 7 results as computed by the PCC approach, it can be seen that the PCC technique has found the majority of the gene spots in this region. The consensus image is generated as per a 66 percent agreement between the layers as indicated. Although this consensus is not perfect it has still found more gene spots in the image region than the traditional approach did.

4.3.4 Strengths and weaknesses

The ability to scale clustering to larger datasets is one of the advantages of the divide and conquer (DAC) approach of the PCC technique. Traditional clustering methods are restricted to the amount of data they can process and as such, clustering techniques can be difficult to implement in practical settings. Another advantage associated with this DAC concept is that of data transparency. Essentially, transparency opens the way for clustering results to be quantified more accurately across an entire surface, irrespective of that surfaces variance. These advantages can be summed up by the results of the synthetic data in Figure 4.16. Traditional clustering approaches have not been able to differentiate between the signal and noise in the lower regions of the image as well as PCC. Indeed, the fuzzy c-means method as applied in the PCC approach has dealt with the background noise element of the input images slightly smoother than that of the k-means method. This smoothness ability is related to the data transparency afforded by PCC.

Unfortunately, a disadvantage of the DAC approach is the added time complexity of the process. As seen in the PCC concept diagram of Figure 4.2, the DAC process splits the surface area up into increasingly larger distinct regions. These multiple regions allow the PCC approach to essentially re-compute a given region slightly and hence account for the intensity differential better. The advantages of this recomputation task outweigh the time complexity delays as the final PCC clustering results are substantially better than those of traditional approaches.

The DAC concept essentially generates ever larger search grids over the input surface. These search grids are then processed one at a time with the individual grid regions slowly increasing in size. This raises the question of what exactly would be an appropriate stopping size. In the current implementation of the PCC algorithm, this stopping size was set to the actual dimensions of the input image. This means the final search grid is a clustering of the full size image logical representatives of layer 0. In fact, this is a "belt and braces" approach as the PCC technique need only extend the search grid to the half way point of the image to render enough image knowledge. As the search grids become larger however their process time is reduced, hence the "belt and braces" stages add more information.

One could argue that as the PCC process does not take advantage of the results as generated by the ITE (see chapter three for details) there is an element of processing that is unnecessary. Remember however that the CMA framework is designed such that there is a slight overlap between the framework's component parts. This overlap helps to increase the robustness of the system (by doing a bit of double checking) while at the same time allowing greater flexibility.

4.4   Summary

This chapter has examined the effects of applying the novel *Pyramidic Contextual Clustering* (PCC) technique to microarray image data. It has shown that through the divide and conquer (DAC) approach the PCC technique was able to cluster a full size microarray image. Coupled with this DAC approach, the technique also renders transparent a clustering method's "internal decision process." Such transparency yields the ability to deal with outlier objects within the image surface more specifically (an area traditional clustering methods fail to cope with).

The clustering result holds sufficient gene spot surface information for later stages of the *Copasetic Microarray Analysis* (CMA) framework to correctly identify and quantify the gene spot regions. The technique presented has shown great potential in the clarification and identification of the gene spot regions and future development work in this area is promising.

Although the results of this PCC approach to clustering data provide improvements over those of more traditional techniques, improvements can still be made. One area that could be improved upon is that associated with the implementation structure. As seen in Figure 4.2, the DAC splitting of the input image region increases the time required to compute the initial time-slice component, with subsequent slices requiring less time. The search grid associated with this DAC approach would gain significantly if it integrated region data from the *Image Transformation Engine* (ITE) results. Such an integration stage would render the ability to focus more specifically on probable gene spot regions, at the expense of reducing the overall PCC result knowledge. Clearly, a carefully designed strategy would be required to take advantage of such an integration stage.

With the completion of the PCC and ITE discussions, the microarray image data preparation stages are now complete. The next chapter of the book focuses on the identification and quantification of the underlying gene spots in the images.

5

Structure Extrapolation II

5.1 Introduction

The research in Chapter 4 identified problems involved when attempting to apply traditional clustering methods to the analysis of full size complementary deoxyribonucleic acid (cDNA) microarray images. These problems included such issues as computational complexity, loss of image information and data blindness. With the application of the *Pyramidic Contextual Clustering* (PCC) and *Image Transformation Engine* (ITE) components to the image data, these issues have been largely resolved. That is to say, computational complexity and data loss issues have been significantly reduced in time and severity respectively. The data blindness (no prior domain knowledge) issue was also followed during the execution of these components, with the individual component results capturing an image's gene spot detail very well. Such detailed gene spot information makes feasible the acquisition of a given image's true gene block structure. Although neither of the ITE or PCC generated results capture the entire gene spots regions in an image (except in the trivial case of a very clean image surface), their individual results can be combined to significantly enhance the gene spot identification process.

As detailed in Section 2.5 a cDNA microarray image consists of several master blocks propagated across its surface, which can also be seen in Figure 5.2. In an ideal situation, these blocks will all be in perfect alignment with each other in the vertical and horizontal directions, both internally to the block and externally across the slide. There will be no rotation or skew with respect to their global positioning. They will not overlap each other or be so close as to confuse their identification at a later stage. Also, internally the meta-blocks will not suffer from any missing gene spots. If the gene spots are missing due to their intensity being weak, they could be obscured by local background.

However, in practice the perfect alignment of blocks within an image is rarely the case. Master blocks are not only misaligned with respect to each other; they also tend to have rotation issues. In the case where there is large rotation in the surface, these master blocks can be difficult to acquire accurately (depending on the identification process involved). Typically though, these rotations tend to be small and as a result are normally insignificant to downstream analysis. The meta-blocks suffer from similar issues, for example, missing gene spots and background contamination of gene spot regions.

The alignment and other related issues highlight the need for designing a method which will correctly address the gene spots within a microarray image. The method should be comparable to some accepted process for the purposes of validation. Although several approaches could render a practical solution for these problems, without some form of validation step it is difficult to quantify the usefulness of a proposed process. The main aim of addressing as it pertains to microarray imagery is to focus on the feature identification problem which can be partitioned into two distinct processes. The first process involves *master block* identification and can be thought of as acquiring the overall layout of the image in question. The second process further analyses these master block locations and acquires the *meta-block* structures or gene spots therein.

The proposed solutions are focused on new techniques specifically designed for the image composition identification tasks. As shown in Figure 5.1, the *Image Layout* (IL) and *Image Structure* (IS) components are executed after the ITE (as was discussed in Chapter 3) has created the default views.* These components complete the Structure Extrapolation stage and can either complement the work of the PCC component or be used independently of it.

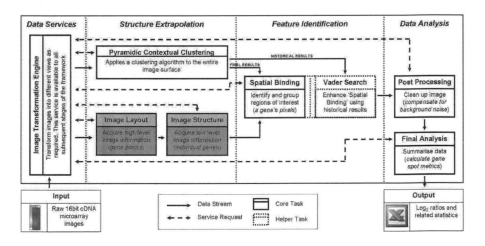

FIGURE 5.1

The CMA framework

---

*As discussed in Section 3.2, the ITE component filters the two or more channel microarray data individually and then merges these channels to create the output image.

The IL component utilizes an ITE generated view of the input image to profile the input images surface. Essentially, the IL component searches for high level repeating constructs within a given surface. Once these constructs (which represent the master block positions in this case) have been adequately identified, the IS component uses similar techniques to define the individual gene spot regions residing within the master blocks. The task of actually classifying these found gene spot regions into signal or noise (foreground or background) will be accomplished by the *Spatial Binding* (SB) component which will be discussed in Chapter 6.

The main focus of this chapter, therefore, aims to build upon the knowledge provided by the ITE and PCC components and define an image's inherent structure in an accurate and directly usable manner. The related gene spot identification problem of feature addressing or gridding as introduced in Section 2.4 will also be examined in this chapter. The chapter shows how correctly identified *master* and *meta-block* locations can be found by relatively straightforward processes when these processes adhere to two important objectives throughout the study:

1) The proposed solutions should determine the master block and meta-block locations as accurately as possible from a microarray image without resorting to operator intervention unless absolutely necessary.

2) The processes should execute in a reasonable amount of time and thus be of low computational time complexity.

It would be advantageous for these processes to function without the need for domain specific knowledge from the outset, with any required domain knowledge garnered from the input imagery as and when required directly (as discussed in Chapter 1). Based on these two objectives, Sections 5.2 and 5.3 present the proposed solutions to the problems as faced when determining master and meta-block structures and the evaluation of the techniques thereof. Section 5.4 concludes the chapter with a summary of the challenges faced when applying these structure identification techniques to the analysis of microarray imagery.

5.2 Image layout - master blocks

In order to determine an image's overall layout or composition, it is beneficial to view the image data such that emphasis is placed upon the overall image structure rather than the raw pixel intensities within. Such modified views of the raw image data have been pre-calculated by the ITE and PCC components (see Chapters 3 and 4 for details). These modified processed images essentially emphasize strong and spatially related signals respectively. This means ambiguities associated with the raw image data (the green noise artifact of the middle image of Figure 5.2 for example) have been reduced in severity.

Figure 5.2 presents a typical input selection for the IL and IS components. The left image is the original test image as used throughout the book and highlights the general image anomalies present. The middle image is a merged representation of the ITE square root and PCC generated results. As stated previously, such a merging of the ITE intensity focused and PCC spatial focused process results give greater enhancement of the image gene spot locations and morphologies. To reiterate, the IL and IS components are not really interested in the shapes of the gene spots at this time, but only their location. The right image in the figure represents the result of a sub-sampling based background estimation process as mentioned in Section 3.2.

FIGURE 5.2
Merged ITE and PCC images. Left: original microarray image; Middle: ITE square root and PCC result; Right: ITE background estimate result

Although the raw microarray image has a great deal of surface variation it is quite difficult to see this variation in raw image (the left image) of the figure. As was shown in Chapters 3 and 4, the modification of these raw surfaces can yield greater knowledge of the underlying structures. The merging of the ITE and PCC generated results have effectively reduced the significance of non-gene spot pixels as seen in the middle image. Note that the white coloration of the gene spots in the middle image are the result of the PCC regions taking greater weighting over the ITE regions.

As already mentioned, both the ITE and PCC components produce results that are not 100 percent perfect, but their merging greatly increases the confidence of a particular gene spot in most cases. Also, note that the outer edge of the green artifact in the top of the middle image surface has been set as gene spot signal in this instance. The right image in the figure represents what the ITE component has determined to be the background or non-gene spot signal. It can be seen from comparing the two artifact regions in image B and C that this artifact noise has been reclassified (by the merge type process) to background rather than gene spot signal. Note these background determined pixel regions generally fall in true background regions or on the outer edge of the gene spot positions. This gene spot edge variation can manifest itself into the diameter and positional inaccuracies, as shown in Section 2.4.

5.2.1 The algorithm

The proposed solution for master block detection is to average the profiles of the slide surface, along the vertical and horizontal directions. The first task of the master block identification process focuses on calculating the vertical positions of the blocks, with the goal being to generate accurate vertical master block demarcation points. These vertical locations can then be examined in more detail in the second task, which determines the horizontal demarcation points. Together these two tasks makeup an accurate location map of the localized master blocks. The pseudo-code for this master block identification process is presented in Algorithm 5.1.

```
 1 Inputs
 2 I : ITE filtered image
 3 C: PCC consensus image
 4
 5 Outputs
 6 Ov: Consensus vertical positions
 7 Oh: Consensus horizontal positions
 8
 9 Algorithm
10 Function (I,C)
11 Process Verticals
12   generate profile of I - Algorithms 5.2 and 5.3
13   normalize profile to remove artifact trends - Algorithm 5.4
14   group profile members by similarity into contiguous areas
15   refine groupings to get consensus positions
16 Process Horizontals
17   generate profile of I
18   normalize profile to remove artifact trends
19   group profile members by similarity into contiguous areas
20   refine groupings to get consensus positions
21 Validate consistency of item 15 and update Ov
22 Validate consistency of item 20 and update Oh
23 End Function (Ov,Oh)
```

Algorithm 5.1: Master Block Identification Pseudo-Code

The profile plots as generated in item 12 of Algorithm 5.1 and shown in Figure 5.5 can be treated as one-dimensional distance plots. These plots were generated with a mean and median operator combination respectively, with the pseudo-code implementation for these operators presented in Algorithm 5.2 and Algorithm 5.3 for the mean and median operator's pseudo-code.

```
1 Input
2 I[y,x] : ITE filtered image, referenced by y/x
3 WinSz: Filter window size centered at pixel
4
5 Output
6 O: Output profile
7
8 Algorithm
9 Function MeanFilter(I,WinSz)
10 For every x and y element in I
11    O= Mean(x-WinSz,y-WinSz x+WinSz,y+WinSz)
12 End For
13 End Function (O)
```

Algorithm 5.2: Mean Filter Pseudo-Code

```
1 Input
2 I[y,x] : ITE filtered image, referenced by y/x
3 WinSz: Filter window size centered at pixel
4
5 Output
6 O: Output profile
7
8 Algorithm
9 Function MedianFilter(I,WinSz)
10 For every x and y element in I
11    O= Median(x-WinSz,y-WinSz x+WinSz,y+WinSz)
12 End For
13 For every x and y element in O
14    O= Mean(x-WinSz,y-WinSz x+WinSz,y+WinSz)
15 End For
16 End Function (O)
```

Algorithm 5.3: Median Filter Pseudo-Code

The master block's periodic features of these one-dimensional profile plots are then calculated by using an autocorrelation (see Appendix D for details) function. For a mathematically random surface, the autocorrelation value will drop quickly from some maximum value to zero (while oscillating about zero). In a purely periodic surface however, the autocorrelation value will drop much slower and will continue to decay to zero (while oscillating about zero). All real surfaces are a combination of these two extreme cases. In cross-correlation, projecting one signal onto another is a means of measuring how much of the second signal is present in the first. This can be used to detect the presence of a known signal or signals within a more complicated signal. The autocorrelation is a special case of this generalized correlation function. Instead of a correlation process between two variables, as in cross-correlation, autocorrelation is performed between two values of the same variable at differing time points. From this periodic information, the master block locations can be extrapolated from the initial positions. These initial grid locations will be tentative at best as they will undoubtedly be misaligned and malformed to some extent. It is important for the following steps to rectify these errors and bring the master block positions into a consistent form.

The second stage, as highlighted in item 13 of Algorithm 5.1, can be seen as a consistency check for each of the grid blocks that have been defined in step one. From the black profiles of the images in Figure 5.3, it can be seen that the profiles lower regions fluctuate across the length of the surface. This is due to the large artifacts as seen in middle images of Figure 5.2. These trends can cause errors in the demarcation points and need to be removed from the profiles (or at least subdued) to render greater accuracy to the master block positions. The pseudo-code implementation for this normalization process can be seen in Algorithm 5.4, which gives IL master block normalization pseudo-code.

```
 1 Inputs
 2 I: raw profile stream (Black line plots in Figure 5.3)
 3
 4 Outputs
 5 O: de-trended stream
 6
 7 Algorithm
 8 Function (I)
 9    Temp=T-smallest value in stream I
10    Pstd=standard deviation of Temp
11    X=find Pstd value in Temp (>=)
12    Update O with the Temp(X) value
13 End Function (O)
```

Algorithm 5.4: Profile Normalization Pseudo-Code

Items 14 and 15 of Algorithm 5.1 examine the peak and valley areas of the profiles in Figure 5.3 to reduce their variability and find accurate profile centers. These peak and valley positions are independently grouped according to any similar differences and are refined until a consensus between the values is found. This consensus data represents a set of consistent block edges that travel the full height or width of the image surface.

Figure 5.3 presents an example final master block result of this IL component process. The white lines in the left and right images represent the master block demarcation points as determined by the IL internal algorithms. The left image shows the initial results of the master block identification process, while the right image shows the same image after the master block splitting process has been applied to the initial demarcation points.

In the left image of Figure 5.3, note how the demarcation algorithm has successfully identified all of the master block regions without the background regions impacting on block accuracy. However, due to slight rotational inconsistencies these blocks could cause a problem for later processing. In order to resolve such an issue, the master block splitting process was designed to demarcate the master blocks into tighter bounds. The results of the splitting

FIGURE 5.3

IL calculated master block positions. Left: Initial master block positions; Right: final master block split positions

process can be seen in the right image of Figure 5.3. Such a tightening of bounding block locations will have advantageous effects upon the analysis of the block structures.

5.2.2 Evaluation

The framework's addressing component splits the master block and meta-block addressing processes into two stages, which produce different equivalent structures to that of GenePix. It is difficult, therefore, to validate the resultant master block or meta-block structures directly as being correct or not. One possible method is to assume that every master block must be of a similar size. However, in practice, there can be large variation within the block positions in a given region (see the bottom right and top right blocks in the right image of Figure 5.3). Alternatively, it could be assumed that every block would have some pre-determined edge profile. This is impractical due to the noise issues raised in Section 3.2, and it is therefore difficult to ensure 100 percent that a gene spot does not reside under this noise element.

Alternatively, the GenePix determined gene spot locations could be used to find the outer edges of the master blocks. Although the IL and IS processes are designed to be fully blind to any prior image knowledge, the use of such domain knowledge for validation purposes does not break this blindness criterion as it is a validation step. Therefore, this GenePix positional information yields an appropriate addressing score for a given master block location. These master blocks move slightly across a surface for a given image and therefore there can be no true location per se. On closer examination it was determined that this validation step is actually built into the IL and IS process mechanisms. Both processes constantly validate their results against themselves. For the IS component to start, the IL component has to have passed all master and meta-blocks as acceptable to regularity conformance and other quality control metrics.

The problem of validation acceptance then becomes one of structure learning, which raises two questions. First, the structures in the image data need to be identified. Second, the variation of process results needs to be analyzed. For most microarray experiments, the gene spots on the slide are divided into equally sized gene blocks which are proportional to the matrix structure printing head of the hardware (see Figure 5.4). This artificial grouping is designed to aid in gene spot identification at later analysis stages.

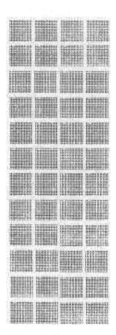
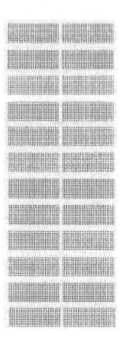

FIGURE 5.4

Example block layout structures. Left: 12×4 structure; Right: 12×2 structure

A cDNA glass slide surface is prepared such that a gene spot could be printed to any part of the upper surface area (see Section 2.4). Ultimately, this ability allows a microarray device to produce a large range of possible gene block layout structures. Examples of the two most common slide structures seen in images during the testing of the addressing techniques are shown in Figure 5.4. Note that both of these layout images were rendered from the GenePix analysis of gene spot positions as determined by a human operator. The images also highlight the misalignment issues of the gene blocks as mentioned previously. The lighter regions in the gene blocks are due to the operator creating smaller gene spot regions for given gene spots.

The microarray layout shown in the left image of Figure 5.4 contains 12 rows by 4 columns. Such an image contains 11520 gene spots divided into 48 blocks yielding 15 rows and 16 columns or 240 genes per block. The layout of the right image of Figure 5.4 separates the gene spots into 12 rows by 2 columns. Meaning the image consists of 9216 gene spots which are divided into 24 blocks giving 12 rows and 32 columns or 384 genes per block.

A total of 122 microarray images were used throughout the addressing evaluation phase, of which five of them could not be correctly addressed without human intervention. These five negative images are not particularly "bad," but they do have an excessive amount of regular high intensity noise. Also, a substantial number of gene spots have been negatively expressed in these images. The combination of this high noise and low signal causes the IL internal algorithms to try and identify regular structures within the noise domain of the image rather than the signal (gene spots). The master block process essentially "learns" the structure of the underlying microarray images. The remainder of this section focuses on several common problem areas seen across the test images to gain a clearer understanding of the results. These problems will typically be related to various repeating forms of noise objects.

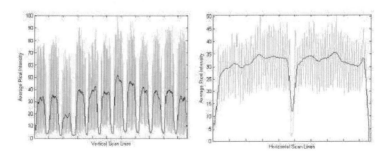

FIGURE 5.5
Aligned vertical and horizontal profile masks. Left: vertical profile; Right: horizontal profile

Figure 5.5 presents the results of this profiling process for the original microarray image (the left image) in Figure 5.2 for the appropriate orientation. The left image of Figure 5.5 represents the vertical positions, as determined by travelling from top to bottom in the surface. The grey line in the image is a plot of the modified surface intensities. The black line is generated by using a combination of sliding average filters, which is a good example of how low pass filters can be applied to improve data quality for human and machine interpretation [152]. The right image of Figure 5.5 shows the horizontal positions, as determined by travelling from left to right in the surface. The grey and black line plots represent the modified and sliding window average results respectively.

From the black line plots of the left image of Figure 5.5, it is relatively easy to distinguish the twelve master rows or verticals that exist in the surface (the peak positions) and the gaps between them (the valley positions). Note also that the bottom edge of the grey line plot in this image fluctuates slightly across the entire surface (this is an indication of the background variability as seen in the middle image of Figure 5.2. The right image of Figure 5.5 shows the results of the same process as applied in the horizontal direction of the image surface. Although this image is classed as relatively "good," there are at least two large noise elements on the surface. One of these noise elements covers the full width and a large part of the height. However, in this case the background pixels are not significant enough to destroy the tracking capability of the process. If there is more noise in an image, there will be less regularity available for tracking purposes. With less regularity the process will rely more heavily on its quality assessment functionality.

Generating the initial plots is a relatively straightforward task; however, these profiles need to be parsed in such a way as to allow the quantification of the master block features. With the human eye, it is easy to see that the vertical and horizontal positions of these master blocks are identified by the black profiles of images of Figure 5.5. But due to the variability of these profiles it is beneficial for a computer system to use a multi-step approach when parsing such data. This multi-step approach is designed to take advantage of any available knowledge for the image. At this point, in the framework this knowledge could have been acquired by the ITE or PCC components. The approach also needs to encompass some form of consistency check in order to aid in block discovery.

Figures 5.6-5.8 represent three example microarray images (the IL slide addressing process) that have been processed by the IL master block component of the framework. The examples are indicative of the various types of noise and other such artifact issues that can cause challenges for the algorithms. The left hand images in Figures 5.6-5.8 show the final master block structure positions. The right hand images are the pixel movement surfaces of the left hand, and give a better indication of the artifact issues the IL algorithms deal with. The middle image of Figure 5.8 shows the blended ITE and PCC result used for this left hand image, which has not helped the structuring process in

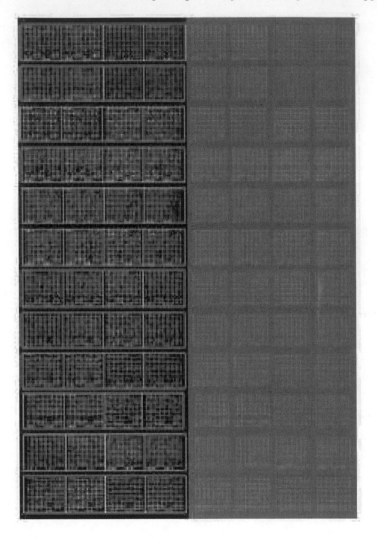

FIGURE 5.6

Example of the IL slide addressing process: the good

this case (this is hinted at by the movement surface on the right being highly irregular).

The left hand image in Figure 5.6 is representative of very clean hybridization images that consist of low regular background noise. From the right hand image, this regular structuring process is clear as there is very little pixel movement between the foreground and background domains. The master block algorithms have had no difficulty in determining the appropriate demarcation points, with the resultant blocks positioned well around the gene spot regions.

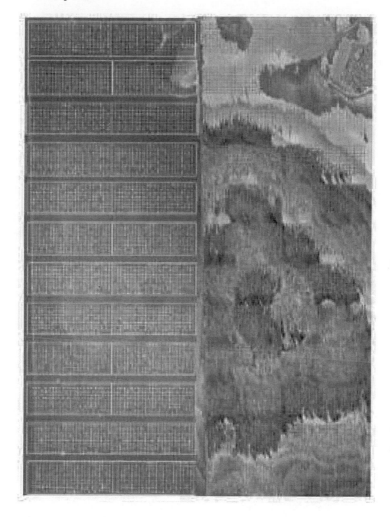

FIGURE 5.7

Example of the IL slide addressing process: the bad

The images in Figure 5.7 represent the most common image quality seen throughout the test dataset. The pixel movement surface in the right of Figure 5.7 is indicative of a highly irregular surface domain. However, the left hand image of Figure 5.7 shows that the IL algorithms have performed well in situations regardless of this background movement content.

The images of Figure 5.8 present one of the five problem microarray images that the master block algorithms failed to correctly structure. As highlighted in Section 5.2, the image composition processes use the multi-view concepts

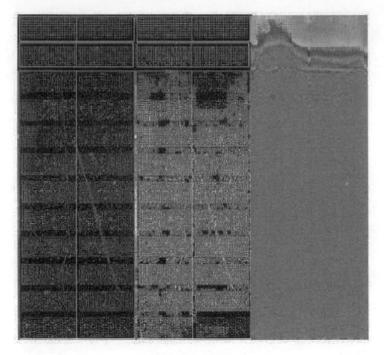

FIGURE 5.8
Example of the IL slide addressing process: the ugly

of the ITE and PCC components to gain insight into a microarray image's internal structure. In the case of the five problem images, this insight has led the IL algorithms down the wrong decision path. The fact that the background movement image on the right is similar to that of its equivalent "bad" version would suggest that the images should be correctly demarcated. However, due to the gene spots being negatively expressed in most cases, the PCC result has become noise enhancing rather than gene spot enhancing as seen in the middle image of Figure 5.8. This reversal of the PCC result has caused the IL addressing algorithm problems with determining a consistent structuring component. We can see the top four blocks of the image have been correctly identified, but critically, due to the PCC weighting element, the remaining blocks have were classified as noise as the consensus of gene spot data was inconsistent. The addressing algorithm has classified the remaining area as one block (the default action), as the assumption (wrong in this case) is the data has different block sizes. This negative spot aspect of the PCC process represents a weakness with the current implementation.

The IL algorithms have performed exceptionally well over 96 percent of the test data. The relatively "good" and "bad" image data fluctuates greatly in quality (see Appendix E for details of all test images), but the IL algorithms have determined a consistent structuring element in most cases. Sometimes the de-trending or normalization of the profile data produced results that were not significantly different from the raw profile data. Given the IL process does not use any external image knowledge (gene spot locations or guide spots for example), the structuring results returned are impressive. The four percent of images that IL failed to determine structure for was due primarily to the PCC algorithms misrepresentation of negatively expressed gene spots as was highlighted in Chapter 4. If the PCC result is not used (no ITE and PCC merging of results) in the IL structure identification process, again, the negatively expressed gene spots (as represented in the ITE result) cause misrepresentation between the foreground (genes) and background (noise) domains in a similar fashion.

5.3 Image structure – meta-blocks

Once the master block positions have been determined, the gene spot positions themselves need to be addressed. At this stage, it is not advantageous to demarcate the gene spots as per the GenePix type process seen in Figure 2.12. Rather, this meta-block demarcation process should instead identify a gene spot's local area (be it foreground or background). In this way, the addressing (of the gene spots) process becomes decoupled from the task of gene spot identification (the gene spots morphology). Such decoupling means any noise artifact present in the determined master block gene regions should not have such a significant impact upon gene spot addressing. Decoupling therefore allows later stages of the framework to be highly optimized to the task of gene spot morphology identification. The remainder of this section discusses the methods created to identify the gene spot regions. As per the IL process, the IS methods are then critically evaluated against a range of relatively "good" and "bad" regions in microarray images in Figure 5.6 and Figure 5.7.

The meta-block process is made up of five sub-processes. The initial process (excluding minor implementation differences) is similar to the master block mechanism, as discussed above. Therefore, the remaining four sub-processes are designed to overcome specific problem areas with respect to gene spot identification, as discussed below.

1) Create Meta-Block - Execute a similar profile method as that of the IL process on an individual master block region.

2) External Block Locations (phase I) - Identify and remove outer locations in the meta-block of 1 which breaks the master block dimension criteria.

3) Internal Block Locations (phase I) - Identify internal grid mismatches from the meta-block of 1 and replace (if appropriate) the offending row or column position information with a consensus of all available meta-blocks.

4) External Block Locations (phase II) - Identify outer grid mismatches from the meta-block of 1 and replace (if appropriate) the offending row or column with a consensus of the meta-blocks that fall along the same row or column.

5) Internal Block Locations (phase II) - In some cases, a meta-block's row or column position might be within an acceptable tolerance with respect to all of the images meta-blocks, when in fact this should not be the case. In these situations it is necessary to rebuild the meta-block of 1 from scratch to find the optimum row and column positions. This process uses all available meta-block knowledge and attempts to rebuild accordingly.

Note that all images presented in this section were designed to enhance the gene spot regions of an image, rather than present a fair blending of the image's two-channel data as previous images. This means any lowly expressed gene spots could be due to the biological experiment itself, or the area in question could be covered by a large artifact. Typically in the case of artifact coverage, the edge detail of the artifact will be classified as a gene spot region. Such an edge classification may seem counter intuitive (as the artifact is not circular, where a gene spot is), but due to the truly blind processing of the IL and IS algorithms however, these edge pixels do resemble gene spots at this point. The five stages of the IS process as outlined above were designed (in part) to overcome such edge classification issues.

5.3.1   Stage one - create meta-block

The meta-block grid is initially calculated by a profiling mechanism similar to the master block process. The main difference in the implementation is the newly defined master block regions are utilized rather than the full image area. In regions of high artifact noise these IS generated profiles will become corrupted, necessitating the need for the multistage functionality seen above. Figure 5.9 presents an example of this initial meta-block grid result for a relatively "good" and "bad" image block region. As stated previously, the images of this section emphasize the gene spots themselves rather than the image surface as a whole and as such will consist of grey scale intensity surfaces. The surface of the left image is a middle block (41) of Figure 5.6, while the right image represents the artifact dominated block (13) of Figure 5.7.

Though the gene spots in the left image in the meta-block are substantially lower in intensity than their left side block members (due to intense background noise in the right), the first pass IS algorithm has correctly identified

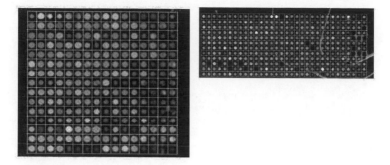

FIGURE 5.9
Example demarcated slide positions. Left: the good; Right: the bad

all gene spots in this region. Note also how the nature of the IS algorithm has centralized the gene spot pixels within the meta-block structure. Even though the region of the right image of Figure 5.9 has a significant background artifact in the right hand side, the IS algorithm has again, correctly identified all gene spots in the region. Including those gene spots that were incorrectly inverted by the PCC process due to their negative expression to begin with (the inverted gene spots on the right). Interestingly, only several of the gene spots in the right hand region were indistinguishable from the artifact and as such were removed.

5.3.2 Stage two - external gene spot locations (phase I)

Stage one defines the initial grid structure for each meta-block as per Figure 5.9. Typically there will still be several issues that need to be resolved before the blocks are in a consistent representative form. Some blocks may have incorrect column and row totals due to one or more outer edges being absent. As the profiling process searches for a consistent set of intensities falling across the surface area, it could quite easily identify artifacts as signals and thus render these incorrect edges. The task of stage two is to identify suspect outer edges and recalculate them according to a consensus from immediate neighbor blocks. The pseudo-code for this stage two task is shown in Algorithm 5.5.

```
 1 Inputs
 2 vgrid: Stage one grids vertical locations
 3 hgrid: Stage one grids horizontal locations
 4 bscore: consistency scores for blocks
 5 mets: IL calculated metrics
 6
 7 Outputs
 8 newVgrid: new vgrid vertical locations
 9 newHgrid: new hgrid horizontal locations
10
```

```
11 Algorithm
12 Function (vgrid,hgrid,bscore)
13    For blk=1 to elements in bscore
14       j=bscore for a given blk
15       vdif=differences for vgrid j
16       hdif=differences for hgrid j
17       vcut=vdif different to mets
18       hcut=hdif different to mets
19       if vcut breaks mets criteria remove appropriate row
20       if hcut breaks mets criteria remove appropriate column
21    End for
22 End Function (newVgrid,newHgrid)
```

Algorithm 5.5: Stage Two Pseudo-Code

If, according to a block's immediate neighbors it has the correct number of rows but too many columns, Algorithm 5.5 will remove the column most unlike those of its neighbors. This removal process is repeated for all of the row and columns within the given block. On the completion of the stage two processes, all of the meta-blocks identified by the IL process will have excess row and column designators removed from their position lists.

Figure 5.10 shows such an occurrence of this column removal process (note the images have been cropped to the regions of interest). The image surface of the figure is from a left hand block (3) of the image in Figure 5.7.

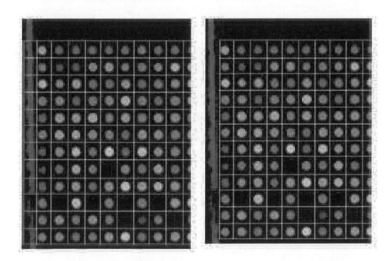

FIGURE 5.10

Example of stage two correction process. Left: initial stage result; Right: stage two result

The artifact remnant running down the left hand side of the image's surface has confused the IS stage one algorithm into addressing the scratch like artifact as a set of gene spots. The application of the external location algorithm has correctly rectified this issue however. Note though that the gene spots in this region exist in a slightly smaller region than their counterparts, this is inconsequential in this case as the demarcation lines do not cut into any of these gene spots. Indeed, any demarcation lines that did cut into the gene spots would be identified as having probable error positions not only by subsequent stages (as will be shown), but also the failsafe stage five algorithms and would be repositioned as appropriate.

5.3.3   Stage three - internal gene spot locations (phase I)

Once the meta-block's external dimension errors are recalculated, the blocks represent partial grid structures as they will typically have missing rows and columns. This sub-process must parse the internal meta-block structures and determine the most probable error row and column positions. The stage considers the outer row and column positions results of stage two as the maximum bounding box positions. As such, this stage is concerned with correcting any internal row and column designator errors. Various metrics were calculated during the master block process. These metrics are reinforced with a meta-block calculation and are used to acquire such information as average column spacing and gene spot distances. The stage's output is an improved set of meta-block layout structures for the slide. The pseudo-code for this stage three task is presented in Algorithm 5.6.

```
 1  Inputs
 2  vgrid: Stage one grids vertical locations
 3  hgrid: Stage one grids horizontal locations
 4  bscore: consistency scores for blocks
 5  mets: IL calculated metrics
 6
 7  Outputs
 8  newVgrid: new vgrid vertical locations
 9  newHgrid: new hgrid horizontal locations
10
11  Algorithm
12  Function (vgrid,hgrid,bscore)
13      For each suspect vgrid
14          calculate intensity distribution for top of block
15          calculate intensity distribution for bottom of block
16          set buildd to direction depending on spread
17          set maxr to maximum gene spots per row
18          For each maxr
19              calculate new vgrid in direction buildd based on
                    gene spot size and nearest neighbor positions
20              if neighbor tolerances are broken, do not modify
                    position
21          End
22      End
```

```
23    For each suspect hgrid
24       calculate intensity distribution for left of block
25       calculate intensity distribution for right of block
26       set buildd to direction depending on spread
27       set maxc to maximum gene spots per column
28       For each maxc
29               calculate new hgrid in direction buildd based on
                     gene spot size and nearest neighbor positions
30               if neighbor tolerances are broken, do not modify
                     position
31       End
32    End
33 End Function (newVgrid,newHgrid)
```

Algorithm 5.6: Stage Three Pseudo-Code

If a meta-block consists of an area that is washed out or contains gene spots that have not hybridized well with the slides local probes, the acquisition of the weak gene spot slide regions can be problematic. The problem can be alleviated with the appropriate use of the IL master block dimension data. At this point, the master block dimension positions are known with high confidence. This means the IS algorithm is aware the region under investigation should be consistent with all other master block regions for this slide. As such, a cross comparison can be made between a given meta-blocks region and all other master blocks to determine valid row and column positions. This comparison must be weighted towards the given block's immediate neighbors due to uncontrollable spatial variations across the slide surface. The correction process is repeated for all of the rows and columns within the given meta-block. When completed, all of the meta-blocks identified by the IL process will have their internal row and column demarcation points brought into line with a consensus view. Such a consensus result is shown in Figure 5.11. The image surface of the figure is from a left hand block (4) of the image in Figure 5.7.

The stage three IS process has fine tuned the addressing positions as determined by stage two. Close examination of image A reveals that in this case, some of the underlying gene spot locations have had their lower regions prematurely cut. This could mean that depending on the state of the gene spot directly below these "cut" regions, the lower gene spot could gain significant background area (as well as the two genes involved being quantified with lower expression than maybe expected).

The stage three processes have recalculated (to finer tolerances) the entire meta-blocks positional data, which has resulted in the gene spots becoming more "central" within their regions. Most of the master block regions by this stage will have correct gene spot bounding positions. The exceptions to this are those blocks that are surrounded by significant recurrent artifact noise or significant noise. Recall from the discussion in Chapter 3 that although the ITE algorithms are able to successfully remove the majority of an artifacts internal noise, edge effects are left intact. The removal of such extreme internal intensities will undoubtedly also remove underlying gene spots. However, the

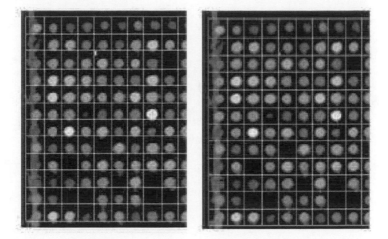

FIGURE 5.11

Example of stage three correction process. Left: stage two result; Right: stage three result

emphasis in the IL and IS processes is placed on regular intensity distributions. This emphasis translates such that the central location of the gene spots is a by-product of regularity itself; this means the gene spots position within the grid area is of no real interest at this time.

5.3.4   Stage four - external gene spot locations (phase II)

Any external meta-block dimension information that broke the master block criteria as determined by the IL process will have been corrected. As the IS stage two sub-processes identified and removed rows or columns falling outside the master block area, stage four identifies and adds any rows or columns missing from the master blocks. Algorithm 5.7 highlights the pseudo-code for this stage four two task.

```
 1 Inputs
 2 vgrid: Stage one grids vertical locations
 3 hgrid: Stage one grids horizontal locations
 4 bscore: consistency scores for blocks
 5 mets: IL calculated metrics
 6 t_struct: IL calculated master block layout
 7
 8
 9 Outputs
10 newVgrid: new vgrid vertical locations
11 newHgrid: new hgrid horizontal locations
12
```

```
13  Algorithm
14  Function (vgrid,hgrid,bscore)
15     # add if too few
16     For each suspect vgrid
17        calculate block dimensions
18        if vgrid < than master blocks consensus
19           For each element of the suspect block
20              set r to appropriate t_struct entry
21              For each r element
22                 Set targetv to starting and ending block
                      elements
23              End
24           set cut to standard deviation of targetv over all
                 t_struct entries
25           if cut = 1
26              update vgrid start element by adding one gene spot
                    top
27           if cut = 2
28              update vgrid end element by adding one gene spot
                    bottom
29           End
30        End
31        update block dimensions
32     End
33     # remove if too many
34     For each suspect vgrid
35        calculate block dimensions
36        if vgrid > than master blocks consensus
37           For each element of the suspect block
38              set r to appropriate t_struct entry
39              For each r element
40                 Set targetv to starting and ending block
                      elements
41              End
42           set cut to standard deviation of targetv over all
                 t_struct entries
43           if cut = 1
44              update vgrid start element by adding one gene spot
                    top
45           if cut = 2
46              update vgrid end element by adding one gene spot
                    bottom
47           End
48        End
49        update block dimensions
50     End
51     # add if too few
52     For each suspect hgrid
53        calculate block dimensions
54        if hgrid < than master blocks consensus
55           For each element of the suspect block
56              set r to appropriate t_struct entry
57              For each r element
58                 Set targetv to starting and ending block
                      elements
59              End
```

```
60          set cut to standard deviation of targetv over all
               t_struct entries
61          if cut = 1
62             update hgrid start element by adding one gene spot
                  left
63          if cut = 2
64             update hgrid end element by adding one gene spot
                  right
65          End
66       End
67       update block dimensions
68    End
69    # remove if too few
70    For each suspect hgrid
71       calculate block dimensions
72       if hgrid < than master blocks consensus
73          For each element of the suspect block
74             set r to appropriate t_struct entry
75             For each r element
76                Set targetv to starting and ending block
                     elements
77             End
78             set cut to standard deviation of targetv over all
                  t_struct entries
79             if cut = 1
80                remove hgrid start element
81             if cut = 2
82                remove hgrid end element
83             End
84          End
85       update block dimensions
86    End
87 End
88 End Function (newVgrid,newHgrid)
```

Algorithm 5.7: Stage Four Pseudo-Code

For example, if there are the correct number of rows in a block and one column is missing, the algorithm can confidently assume that this column should have been on an outer edge (as the internal structure has been verified in IS stage three). Here, a column will be placed at the left and right edges of the block. The column which best matches the algorithms slide spot metrics and thus resembles an underlying block structure more accurately (and does not break any dimension criterion garnered from previous blocks) will be kept. An example of this addition to outer edge blocks is highlighted in Figures 5.12-5.13. The image surface of Figure 5.12 is a test microarray image that has slight washing artifact interfering with the gene spots in the bottom left hand region of the array, while images in Figure 5.13 and Figure 5.14 are from the highlighted block (12) of Figure 5.12. Other than the artifact region in Figure 5.12 (and the significant background border noise), the image surface itself is generally clean. Indeed, only the neighbors of the highlighted block and the slightly darker blocks in the middle-right of the surface are lacking in gene spot signal.

FIGURE 5.12
Example of stage four correction process: original microarray surface

Closer examination of the images in Figure 5.13 and Figure 5.14 presents this "missing column" problem as described earlier. The column on the left hand side has failed to be classified as containing gene spots (only five gene spots can clearly be seen in this region).

5.3.5   Stage five - internal gene spot locations (phase II)

In most cases the slides identified master blocks will now be correctly defined with the rows and columns demarcating the central valley points for the gene spots. However, a situation can arise where these demarcation points can intersect the gene spots even though they adhere to a global consensus structure for the slide's surface. Therefore, a final check process is required to verify that the demarcation points in a given meta-block fall within acceptable positions. The check process is essentially a variation of the IS stage three implementation, with the difference being that the build direction processes are calculated in a different manner. As such, the pseudo-code presented in Algorithm 5.8 shows only the differences between the IS process three and five algorithms.

```
 1 Inputs
 2 vgrid: Stage one grids vertical locations
 3 hgrid: Stage one grids horizontal locations
 4 bscore: consistency scores for blocks
 5 mets: IL calculated metrics
 6
 7 Outputs
 8 newVgrid: new vgrid vertical locations
 9 newHgrid: new hgrid horizontal locations
10
11 Algorithm
12 Function (vgrid,hgrid,bscore)
13 .
```

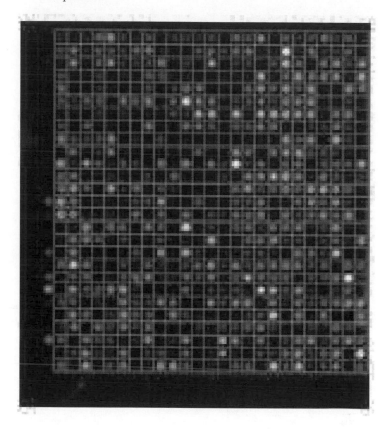

FIGURE 5.13

Example of stage four correction process: stage three result

```
14    #remove suspect block rows and columns with acceptable
          locations
15    .
16    .
17    set buildd to direction depending on spread
18    set maxc to maximum gene spots per column
19    For each maxc
20       calculate new vgrid in direction buildd based on gene
             spot size
21       if neighbor tolerances are broken , do not modify
             position
22    End
23    .
24    .
25    set buildd to direction depending on spread
26    set maxc to maximum gene spots per column
```

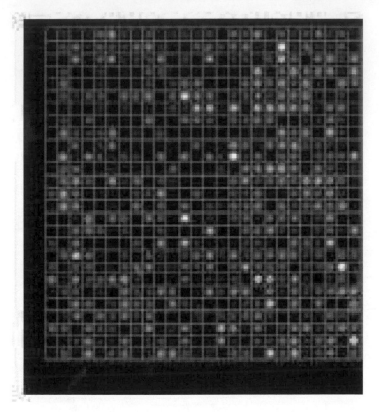

FIGURE 5.14

Example of stage four correction process: stage four result

```
27    For each maxc
28        calculate new hgrid in direction buildd based on gene
              spot size
29        if neighbor tolerances are broken, do not modify
              position
30    End
31    .
32 End Function (newVgrid,newHgrid)
```

Algorithm 5.8: Stage Four and Five Differences Pseudo-Code

Key to the success of the IS techniques is the underlying redundancy between the sub-process stages. Each of these sub-processes overlaps to some extent the others, as this reinforces the consensus view of the image's surface. However, the techniques employed in the IS processes are not foolproof. There are occasions where the IS algorithms are drawn to structure in the image that does not represent the true blocking layout. For these situations a manual analysis typically finds that the image under investigation is covered in substantial systematic noise elements.

5.4 Summary

The culmination of challenges as discussed in the previous chapters (large volumes of data, drastically varying levels of noise and structural layouts) have largely been overcome at this point by using a combination of techniques to provide what appears to be a viable candidate for a fully automated system. The underlying structure of 96 percent of the slides tested were discovered successfully, with the remaining failures attributed either to extreme systematic noise or minor problems that should be resolved in the near future.

This chapter investigated the issues associated with the addressing stage of the *Copasetic Microarray Analysis* (CMA) framework. It was shown that an initially simple concept applied correctly with the appropriate helper functions could correctly address a microarray's gene spots over a range of very different qualities. It was demonstrated that the process results were of appropriate quality for downstream analysis but there is still room for improvement. One possible direction for further studies as highlighted by the *Image Layout* (IL) and *Image Structure* (IS) components of the framework could be along the lines of investigating evolutionary type processes into the IL and IS stages. However, such a concept would need to know what the given gene spot should look like, which under the current implementation is not strictly known at this point. Also a critical aspect of the IL and IS algorithmic successes are tied to the *Image Transformation Engine* (ITE) process results. The cleaner this ITE result, the better the gene spot identification process will be (see the Chapter 3 summary for possible ITE improvement details).

The ITE component is the only process in the current implementation that has access to the raw microarray image. Also, the ITE merges the two or more channels of a microarray experiment into one composite image. In theory these two or more channels should be nearly identical in structure and gene spot positioning.

The experiments in this chapter showed that blind processing of a microarray slide is possible. As the noise became significant, the process failed on its own, however with the correct algorithmic mechanisms human intervention was kept to a minimum. Most current processes of microarray addressing focus on a process that relies on either human intervention throughout or so called guide gene spots in the imagery. In the next chapter these meta-block locations will be utilized to create the gene spot profiles, and will further be analyzed in order to take advantage of the analysis provided by the *Pyramidic Contextual Clustering* (PCC) histories.

6

Feature Identification I

6.1 Introduction

Chapter 5 discussed the challenges faced during the blind acquisition of a microarray's internal composition. The acquisition phase was designed to accurately locate the gene spot master blocks and the individual gene positions within the master block regions. It was shown that the processes involved were able to overcome a range of technical difficulties (scratches, fingerprints, etc) and produce gene spot addressing results appropriate for the surface in question. Once the gene spot regions have been located, the analysis task becomes one of feature identification. The challenge for this feature identification process is to correctly identify the gene spot morphology within the given grid location as determined by the *Image Structure* (IS) component.

All of the necessary pieces are now in place to identify a given gene spot's morphology. This identification task is encapsulated in the *Spatial Binding* (SB) component of the *Copasetic Microarray Analysis* (CMA) framework. As per previous framework components, the SB process should only use knowledge gained directly by a framework component. This means the *Image Transformation Engine* (ITE) and the *Pyramidic Contextual Clustering* (PCC) component results can be used to gain insight into a given gene's location if required. As already seen in the Chapter 5 discussions, the ITE and PCC processes can be extremely powerful if used correctly. The SB component uses the PCC information for a given gene spot location and, essentially, looks back over the time-slices (or PCC history layers) at this position. Although the initial one or two time steps of PCC are littered with numerous noise artifacts, subsequent layers refine the gene spots themselves. SB essentially stitches these partial gene spot pixel locations together (with some restrictions) to create the full gene spot. The final gene spot members are then used to quantify the gene's expression. These final gene members are also spatially compared to results as determined by a human operator's usage of the GenePix package for validation purposes.

With the completion of the Structure Extrapolation stages, the investigation focuses on finding accurate gene spot surfaces. SB examines every *region of interest* (ROI) as determined by the IS process (these regions should contain one gene spot only as determined by IS). The SB process therefore examines these regions at the pixel level to determine the probable location of a gene spot. Once a gene spot's morphology has been determined as accurately as possible, the gene's expression level is quantified with greater confidence.

This chapter aims to explain how these ROI regions are broken up into their foreground (the gene spot) and background elements. In the context of the CMA framework, the process of gene spot identification is encapsulated in the SB component, as shown in Figure 6.1.

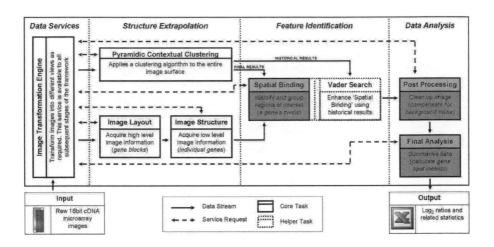

FIGURE 6.1
The CMA framework

The process of feature identification combines knowledge of the gene spot positions with their intensity distributions as learnt from the results of previous stages, which highlights three important objectives for the chapter's investigation.

1) The proposed SB solution should be restricted to the "PCC consensus layer" ROI dimensions at this time to comply with time restrictions as required. An alternative approach will be highlighted in Section 6.5 which generates more accurate bindings (potentially) at the cost of greater time complexity.

2) The process should execute in a reasonable amount of time and, therefore, be of low time complexity.

3) Within the ROI, the potential gene spot morphology should be coerced by some means - an estimate template (for example) - to increase confidence in gene spot pixel member selection.

Based on the three objectives, Section 6.2 presents a broad overview of the proposed solution to the feature identification process. An evaluation of the feature identification process is discussed in Section 6.3 against real-world data, with the CMA framework evaluated as a whole in Section 6.4. The chapter concludes with a summary of the issues faced when identifying the features of a microarray image in Section 6.5.

6.2 Spatial binding

6.2.1 Pyramidic contextual clustering - revisited

As mentioned earlier, the SB process utilizes the PCC generated history results to help ascertain a given gene spot's shape. As such the key elements of the PCC technique are re-iterated here for purposes of flow. The PCC component can be thought of as providing "time travel" like functionality into a clustering method's internal decision process. For the SB process, these "time slices" represent the clustering results for a given gene spot over the increasing meta-grid granularities schema (see Chapter 4 for greater detail). Figures 6.2-6.3 highlight the important aspects of this PCC approach as they pertain to the gene spot identification task, which show the pyramidic contextual clustering process in a nutshell. The image in Figure 6.2 re-iterates the concept of the PCC surface analysis process. The left hand sub-image of Figure 6.2 is the original microarray surface as used throughout the book, the square gene spot region in this image is used at a later stage of examination. The cascaded surface images in the right of Figure 6.2 represent the 13 "time slices" generated by the PCC concept of the center box. The image in Figure 6.3 is a blown up surface plot of the consensus image as generated by amalgamating the 13 individual "time slices" of the image in Figure 6.3.

From Figure 6.3 the bulk of the gene spots in the cascaded images have been retained in this consensus result. The SB process essentially traverses these individual layers while limiting its view point to one gene spot region (the square in Figure 6.2) at a time. In this way, snapshots of the individual gene spot regions are generated such that the *Vader Search* (VS) functionality of SB can determine gene spot pixel classifications more accurately.

6.2.2 The method

The biggest challenge faced by the SB process is that of determining an accurate bind for the gene spot under investigation. Packages such as GenePix (see Section 2.4 for details) for example sidestep these identification accuracy issues algorithmically by requiring the package operator determine the gene spot bounding pixels. This so-called "passing the buck" is frowned upon here

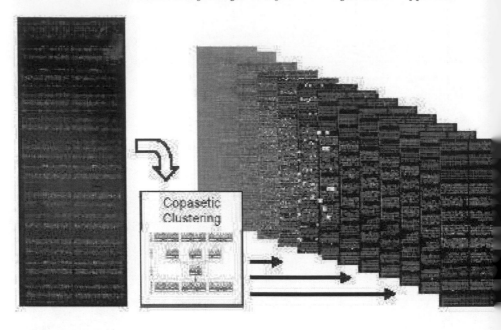

FIGURE 6.2
Pyramidic contextual clustering layer

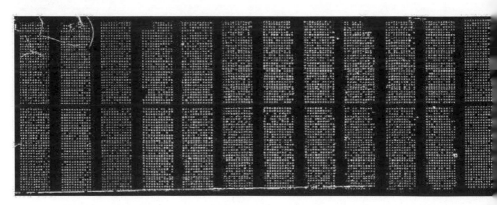

FIGURE 6.3
Pyramidic contextual clustering consensus

as one of the overall aims of the CMA framework is to process the imagery in a truly blind automatic manner. As highlighted in Figure 6.1, the SB process uses the gene spot profile (for the current gene) as generated from the PCC component during the identification stage. Due to the nature of the clustering process (and the way in which the PCC technique executes) this profile will typically represent the gene spot itself (be it a full or partial representation).

One way in which a gene spot could be examined would be to execute a search across the gene's PCC generated profile surface. Such a search would involve tracking the perimeter of the PCC result as well as the internal pixel positions for other similar pixel intensities within the ROI. Greater weighting would be given to the PCC embedded pixels (as the PCC result is biased to the gene spots to begin with), with the returned search result limited to the probable gene spot diameter (as determined by the IS process). The internal perimeter search would highlight pixels that are probably "lost" during the PCC generation step (maybe they were grouped with local background). These lost pixels could then be compared against the profile members, with similar pixels re-grouped with the underlying profile members.

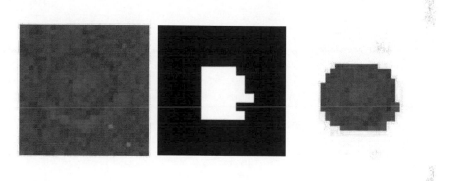

FIGURE 6.4
Example of PCC based perimeter search process. Left: Raw gene spot region; Middle: PCC profile; Right: perimeter search result

Figure 6.4 highlights the central concept of this perimeter search process. The left image is a gene spot from the original surface of Figure 6.2. This gene spot was chosen specifically as it was negatively expressed and as such, exists below the local background region. The bivalence surface of the middle image is the PCC determined gene spot profile region, where the white pixels represent actual pixels of interest in this case. Image surface of the right image is a result of applying the aforementioned perimeter search into the region of the left image.

As can be seen in the middle image of Figure 6.4, the PCC generated profile does not in fact represent the gene spot accurately. However, this PCC result has correctly located the likely gene spot from the region of the left image of Figure 6.4. Therefore executing the internal perimeter search on the file of the middle image of Figure 6.4 (with appropriate caveats as mentioned previously) returns the embedded result of the right image of Figure 6.4. Although the surface profile of the right image of Figure 6.4 is not a perfect gene spot match when compared to the actual surface in the left image, it is substantially better than the result yielded with the raw PCC process. Importantly, remember that the PCC process was designed to find probable gene spots in the image, not accurately determine gene spot morphology. Indeed, any accurate gene spot PCC profiles are considered bonus results and should therefore not be relied upon initially.

A problem with this perimeter search process is realized when more than one object or gene spot is returned in the PCC result. The PCC result is essentially designated the tentative gene spot location; two or more objects bring this designation into doubt. However using ideas generated by this ideal perimeter process, a valid compromise can be found.

This new perimeter process should not rely simply on expanding the PCC profile of above as multiple objects can cause problems. Rather, this new process should expand upon several techniques which, when combined, produce appropriate gene spot profiles. The SB process can be thought of as a grouping together of the pixels that make up a gene spot. The resultant grouping can then be summarized into an appropriate statistic for the given gene spot at a later stage. SB takes ideas generated from precursor research and traditional concepts to determine as accurate a mapping as possible of the gene spot in question.

Rather than using the raw microarray image data initially,* the process uses the imagery generated by the PCC component. As discussed in Chapter 4, the PCC set of images represents a spatially orientated view of the microarray surface focused at suspected gene spot positions. Although this PCC image set is not perfect in its gene spot positional information, there is enough information to render a good gene spot estimate template (ET), and therefore the gene spots center position. The ET structure represents the typical morphology for the known gene spots in the image. SB uses both the ET gene spot center positions and the PCC results to search for and combine groups of pixels that fall within close proximity to each other.

The SB process successfully stitches the majority of genes that are partially or well defined (by PCC) back together, with the information gained reinforcing the ET for a typical gene spot. This ET gene spot structure helps the identification process as the results of the PCC component (and those of clustering in general) can never guarantee that all pixels belonging to a gene spot (in this case) will be identified correctly. Indeed, the ET helps gene spots with a particularly low intensity or that are missing completely from the PCC result to be in recovered as it can be thought of as a global estimate. Such

---

*The raw image data is littered with noise elements that could render the gene spot invisible to begin with.

a recovery process positions the ET in the expected gene spot location and re-examines the region. This re-examination process is encapsulated in the VS helper function of SB. VS trawls the gene spot region of the PCC "time slice" layers (as per the square region of Figure 6.2). VS looks specifically for the changing allegiances of the region's pixels through the time slices, and as such, helps the SB component to accurately define the gene spots. The results of this SB binding process are then stored as the given gene spots members list and are referenced in the final quantification process.

FIGURE 6.5
Example of SB's VS time slice search process. First: gene spot region; Second: overlays; Third: initial pass; Fourth: pass 2 $\cdots$; Fifth: final pass

Figure 6.5 presents a concept diagram for this SB perimeter process. The first image of the figure is a gene spot region from Figure 6.2. Note that the dark band travelling through the central region of the gene spot will have affected the PCC result process for this gene spot. The second image presents the same gene spot region along with three system-generated overlays. The overlay is the ET binding ring as calculated for the whole microarray image surface of Figure 6.3. This binding ring essentially encases the outer edge morphology of gene spot candidates in the PCC consensus image surface. Background noise artifacts will heavily corrupt a gene spot that falls outside of this ring. The ring overlay is representative of the gene spots shape as determined by the SB process and should be equivalent to the human based GenePix identification process result (see Section 2.4 for details). The green ring was generated by a *generalized Hough transform* (GHT) [129] (see Appendix D for details) backup process designed to give an indication of the region's gene spot via a different (non PCC or ET) based approach. Importantly, note how close the ring centers are for this surface region. The third, fourth and fifth images of the figure present a conceptualized workflow of the SB analysis process. From the initial binding of surfaces of the second image and PCC in the third image, through multiple iterations of concentric circles in the fourth image, the final gene spot region is calculated in the fifth image.

The first image in Figure 6.5 has two distinct regions that will be separated by the PCC process. Such a separation can be seen in the third image of the figure (an embedded version of the PCC profile and the first image) where the central dark band region (along with some other pixels) was classified as background by PCC. Using the various ring constructs as mentioned, and starting from their central positions, expanding concentric circles are thrown around the center pixel. Within these circles (as shown in the fourth image), the underlying pixel positions are compared to their PCC classifications, pixels designated as foreground members by PCC are added here as required. Pixels not allocated to the foreground by PCC are analyzed by VS to determine their time slice history.

Whereas the PCC consensus can be thought of as being similar to this VS task, the VS process analyses all PCC layers. The difference being that VS is capable of tracking the shifting allegiances of the pixels in the initial background classified pixels (the early layers). Put another way, if a gene spot pixel was classified as background in an early PCC layer (they are typically), and is then re-classified as foreground it should have been foreground initially and vice versa. Such intimate pixel tracking capability sees the rebuilding of weak and partial gene spot regions to accurate shapes (the fourth image).

Importantly, note that the GHT ring generation process is meant to be similar in goal to the ET ring process. Whereas the ET ring is calculated with respect to all known gene spots in the image however, the GHT ring only takes the gene spot region under examination into account. As seen in the second image in Figure 6.5, the GHT approach to a probable gene spot location within a given region is affected by the regions background noise. As such, this noise element degrades the GHT accuracy somewhat. As previously stated however, the GHT approach is used for guidance during the allegiance phase of gene spot pixel acquisition, with greater priority placed on the ET ring.

Results as generated by an actual SB process over a typical gene spot region are highlighted in Figure 6.6. The left image presents the individual "time slice" layers as viewed for a specific gene spot (red square of Figure 6.2). While the right image is the consensus view of the same gene spot region with the initial (as applied to the original surface of the left image) GHT and SB rings present.

From the left image of Figure 6.6 (a gene spot residing within an artifact region), it can be seen that the initial PCC layers produce various initial results where the pixels are spatially related, but are not consistent throughout the full image surface. The image highlights perimeter pixels (blue ring) that are defined as the gene spot area in each layer. A consensus of these layers is used to help in the identification of the final gene spot (step 3 in Figure 6.5). In the right image, the regions data is coalescing to the gene spots region as the PCC layers granularity parameter changes. As the layer number increases consistent pixel data positions will become stronger due to allegiance effects. These effects can be seen clearly in the right image of Figure 6.6 with the stronger hues depicting regions that are more consistent over the multiple layers.

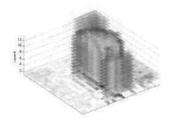 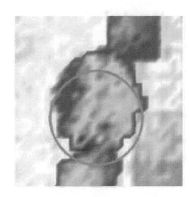

FIGURE 6.6
Example SB gene spot identification result. Left: PCC time slice layers for gene spot region; Right: consensus VS PCC time slice layers in the left image

6.2.2.1   Generating a gene spot's estimate

The next task is to determine where exactly in the IS region the gene spot actually resides. From the results generated by the IS components the SB process is aware of the tentative gene spot locations across the microarray image. With this location information, it is possible to determine (from the strongly expressed gene spots initially, with subsequent rebuilt gene spots added as found) a gene spot's typical morphology and its central location. This morphological and center positional information in turn render the ability to determine the given gene spots region location. Also, as the IS component calculates the diameter of the gene spots (globally), an idealized boundary for a given gene spot is estimated.

The left image of Figure 6.7 presents an overview of the results generated by the IL and IS components. The top close-up is a section of this overview, detailing the gene spot morphologies found in a particular master block region. The bottom close-up section highlights the details of an individual gene spot from this master block. The right image shows an example of such a gene spot estimate as generated in this manner.

For the left image of Figure 6.7 in this case, the gene spot resides on the outer edge of an artifact region and the artifact edge has also been identified as a possible gene spot region by PCC. To create an estimate of the typical gene spot morphology in the image, the individual grid regions of the master blocks (as defined by the IS component) are traversed. The contents of these regions are then amalgamated to create a final ET global gene spot shape. As the ET result takes into account the shape of all known (PCC highlighted) gene spots in the image, the shape is not specific to a particular gene spot. The returned shape is however a good indicator of the outer edge of the genes.

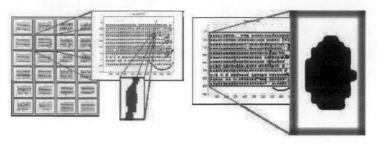

FIGURE 6.7
Example estimate template generation. Left: image layout and structure;
Right: estimate gene spot template

6.2.2.2 Recovering weak genes

Once the PCC layers, the IL and IS inspired gene spots estimates are in
place, the task of recovering weak gene spots from the surface begins. Figure
6.8 highlights the main steps involved in this gene spot rebuilding process.

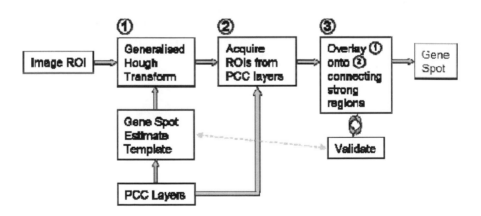

FIGURE 6.8
Pipeline of spatial binding process

The image ROI as seen in the first block of the pipeline represents the gene
spot region as determined by the IS component. Specifically, the ROI is culled
from the PCC consensus layer using the aforementioned IS calculated region's

dimension information. A GHT is then applied to this ROI (step 1 blocks), with internal parameters set as determined by the IS process, the results of which give a good indication of gene spot location within the ROI.

As will be shown later, although the GHT method is robust to noise it can still be misguided in its calculations. Therefore, this GHT shape determination method is backed up by the ET technique. This estimate technique takes a consensus of all available gene spot morphologies found by the PCC component. Although the ET process does not typically generate profiles as accurately as the GHT method, the resultant is a good indicator of gene spot shape and general proximity. When these two methods are used together, the gene spot is rendered with greater confidence.

Once the GHT and ET techniques have been applied to the ROI as required, the step 2 block is executed. Here, all available PCC layers for the image associated with the ROI are traversed. During this traversal process, the ROI itself is acquired from every layer and stored appropriately. As all of the required pieces are in place, the task of rebuilding the suspected gene spot can begin. The rebuild process takes the GHT analysis result for the ROI and overlays it onto the acquired PCC layer sequence one slice or frame at a time. The pixels contained in or falling on the edge of the GHT binding ring and active in the PCC layer are tentatively classified as valid gene spot members.

As the SB process passes through the PCC layer ROI, the VS component executes a pixel/slice re-classification task. This task examines which of the "valid" pixels from above are changing their allegiance (foreground to background or vice versa) through the PCC "time slices." For example, if a central pixel is classified as foreground initially, but at a later time, it changes allegiance to become a background member, it will be examined by the VS task. In this way, starting from the central area of the suspected gene spot, the information generated by the PCC process is extended on an individual gene spot pixel basis. This re-classification task is repeated with the gene spot ET process used as the binding ring. It is important to note that the gene spot GHT identification process acts as a backup to that of the ET process (there will be a slight disparity between the two perimeters, but the majority of the pixels highlighted will be the same).

The final result from this pipeline process of Figure 6.8 will be the list of probable pixels from the ROI that define the gene spot. Algorithm 6.1 presents pseudo-code for the GHT and VS gene spot location and allegiance tracking processes respectively.

```
1  Input
2  I[y,x]  : Raw image, referenced by y/x
3
4  Output
5  O: Output image
6
7  Algorithm
8  Function SpatialBinding(I)
9      Get PCC consensus layer image
```

```
10    Get PCC layers for image
11    For every grid region in image
12       F=foreground pixels
13       ET=average gene spot morphology
14    End for
15    For every grid region in image
16       [y,x,acc]=GHT(ROI,dia,size)
17       Determine ROI through all PCC layers
18       From center of ROI as determined by GHT
19          connect strong pixel elements together
20       Validate chosen pixels via GHT, ET and SB perimeter
21       Write validated pixel list to file
22    End for
23 End Function (0)
```

Algorithm 6.1: Spatial Binding/Vader Search Pseudo-Code

After acquiring the appropriate ITE view and all PCC layers for the image, the process first generates the ET structure for known gene spots. This global average is then used along with the current gene spot estimate (via GHT and iterated SB process) to determine the gene spot morphology.

6.3 Evaluation of feature identification

This section introduces three example gene spots from microarray images that are used to evaluate the performance of the SB technique. The three gene spots shown in Figures 6.9-6.10 were chosen, as they are good representatives of the type of challenges faced by the SB processes. Emphasis is placed upon particularly poor quality regions of the imagery, with these region results discussed in detail with respect to the SB sub-processes.

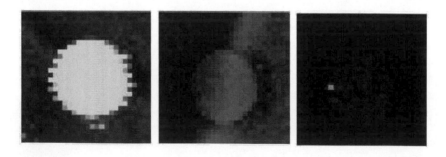

FIGURE 6.9
Three original gene spots. Left: the good; Middle: the bad; Right: the invisible

FIGURE 6.10

Associated movement surfaces of the gene spots in Figure 6.9. Left: saturated; Middle: ambiguous; Right: not so invisible

The gene spot in the left image in Figure 6.9 was overexpressed during the hybridization process of the experiment run. This overexpression has had the affect of saturating the scanner devices processing. Recall that scanners are capable of uniquely recording 2^{16} unique values. The gene spot pixel values in the left image in Figure 6.9 have been capped at this maximum permissible value. Note also the disruption in laser frequencies in the gene spot region as evidenced by the pure red and green pixel elements and the alternating stepping structure of the genes outer edge. Clearly, this gene spot can be separated from the local background region with little difficulty. The surface of the middle image in Figure 6.9 is representative of a typical gene spot straddling an artifact region. The gene spots profile can be seen clearly on the right hand side, but not so well on the left. This gene spot will cause issues to traditional separation processes due to it similarity with background regions. The gene spot of the right image in Figure 6.9 has such low intensity that the immediate neighbors have overshadowed it. Gene spots of this nature can be very difficult to separate from their background. The image surfaces in the bottom row of the figure show the pixel movement transform surfaces (see Chapter 3 for details) for the above-mentioned regions. Unsurprisingly, the movement surface of the left image in Figure 6.10 shows that the gene spot surface profile for the left image in Figure 6.9 is highly consistent and smooth, with some background variation. The middle image in Figure 6.10 is more ambiguous as the gene spot pixels clearly fluctuate quite rapidly over the gene's surface. Note also that the background is relatively static in the image. The movement surface of the right image in Figure 6.10 shows, surprisingly perhaps, that there is in fact a gene spot in this region with consistent or regular foreground and background intensities.

It can be seen in the left image in Figure 6.9 that this gene spot's surface has a stripped effect. Typically, the striping problem is associated with the scanning technology used in the digital generation stage (see Appendix A for details) or an excessive hybridization-binding phase. In the case of the former, this is typically due to the scanner's laser or photomultiplier (PMT) parameters not being set correctly for these types of gene spot. For the latter, the binding phase rendered a hybridization that sits close in intensity level to the scanner's maximum element value (2^{16}). In either case, the resulting feedback value from the lasers is capped at this maximum value. These effects can be reduced with greater care and attention being focused on the pre-scanning stage of the digital generation step.

In the middle image in Figure 6.9 there is ambiguity between the foreground and background elements (especially on the left hand side of the gene spot). Issues of this nature (inconsistent surface intensity for example) can cause traditional techniques (clustering) difficulties, but the SB process functionalities are hybrid in nature (and have several backup tasks) and can overcome these ambiguous separation decisions relatively easier. The right image in Figure 6.9 represents the other (as opposed to the saturated surface of the left image in Figure 6.9) extreme that can be seen in a microarray's surface. The region does actually contain a gene spot but due to the extremely low expression of this gene, the immediate neighboring gene spots have masked the morphology.

6.3.1 Finding a gene spot's location and morphology

As stated previously, in order to rebuild a gene spot stored within the appropriate IS region, the first step is to identify the genes spot's probable morphology and location within the region itself. In some situations, this process is relatively straightforward due to the large intensity differences between the two surfaces (the left image in Figure 6.9). However, in a more typical gene spot region (the middle and right images in Figure 6.9) this identification task becomes less trivial. In a similar vein as that of the ITE process of a multi-view approach to data cleaning, the SB component utilizes several approaches in this gene spot identification task.

The GHT technique is utilized here as one such multi-view task, which offers an independent (gene spot specific) approach to identify the likely location of said gene spot within the IS region. Essentially the GHT accumulator function examines the gene spot region looking for an object with a diameter of x, where x is a refinement of the IS determined diameter metric. In an ideal situation, where the gene spot is substantially greater in intensity than its local background, this accumulator function (given the correct diameter) will pinpoint the gene spot. Any slight variation in the gene spot size or background noise to the ideal will result in errors for the gene spot center. As a backup process however, this gene spot location estimate is adequate for the task.

The accumulator (see Appendix D for details) results are presented in Figures 6.11-6.12 for the sample gene spots of the three images in Figure 6.9. Figure 6.11 is the accumulator results with the ring highlighting the center position. Figure 6.12 highlights the tracking ability of the accumulator when overlaid directly onto the combined "time slice" views of the relevant PCC layer regions.

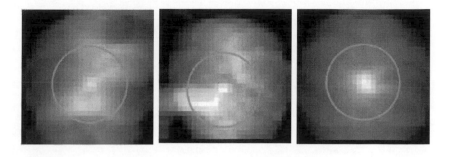

FIGURE 6.11
Example gene spots GHT accumulator results. Left: the good; Middle: the bad; Right: the ugly

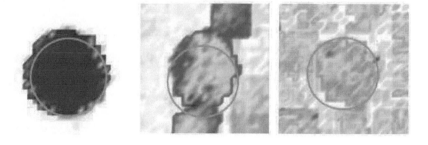

FIGURE 6.12
Example gene spots GHT accumulator results. Left: ideal tracking; middle: pretty close; Right: excellent

From the left surface images in Figure 6.11 and Figure 6.12, it is clear that the saturated gene spot has been identified fairly accurately within the region. The relatively smooth surface of the left image is indicative of a fairly consistent foreground/background intensity spread in the region, with the GHT emphasizing a center area of the region as the gene spot center. This indication is confirmed in the left image in Figure 6.12 where the same green location ring of the GHT process has been overlaid onto the PCC generated consensus region for the gene spot. Note also in the left image in Figure 6.12 that there is very little background noise, which further indicates the PCC result focused on the gene spot rather than the background.

In the middle image in Figure 6.11, some of the background noise had been classified as possible gene spot members. These members are relatively large (top and bottom square regions) and their intensity similarity to the gene spot surface has caused problems for the GHT accumulator. However, as mentioned previously the GHT process requires a size parameter to be set, which renders the ability to look for specific patterns (diameters more accurately) in the data. This parameter is calculated as part of the IS addressing process and thus can be passed to the GHT task. The size parameter has helped guide the GHT process in this case, away from the two square artifact areas of the image (the artifacts do not fit the size criteria sufficiently), and identified the more likely off-central pixel as being the gene spot center. Although this bind fitting is not perfect in this case (the high intensity below the gene spot has attracted the GHT ring), it is adequate for the VS process as will be discussed presently.

The result as generated for the right image in Figure 6.12 is perhaps particularly unexpected. From the original surface area of the right image in Figure 6.9, the region did not have obvious gene spot pixels (excluding the highly expressed singular one). As hinted at by the movement image however, there was quite clearly "something" in the surface region. Indeed, in this case the GHT has identified what looks to be a close approximation of the true gene spot location. The GHT identification process is utilized primarily for the robustness to noise variation that it offers the SB component. Although the GHT process is robust at identifying probable gene spot locations within noisy gene spot regions, it is not perfect in its identification task. Therefore, further processes are required to help reinforce this initial identification result.

6.3.2   Recovering weak genes

Once the GHT and ET processes have tentatively identified the gene spot locations within the IS determined ROI, the VS functionality is executed. The premise of the VS technique is that the pixel members associated with a given gene spot would (through the PCC "time slices") have been classified as foreground at some point in their history. The task for the VS process then is to traverse these "time slices" and reassess the allegiance of the probable gene spot member pixels. This re-assessment task is bound by the GHT and ET regions as the gene spot under investigation is guaranteed to sit somewhere under these regions. The gene spot therefore will be bound by the ET global result, with the GHT bound result offering a good starting location for the VS process to search for the gene spot itself.

The images of Figure 6.13 represent the initial PCC layer surface for the reference gene spots regions as shown in Figures 6.9-6.10 earlier. The white squares in the three regions represent pixels which were classified as foreground elements at the specified "time slice" layer. The overlay is the ET binding ring as calculated for the whole microarray image surface. This binding ring essentially encases the outer edge morphology of gene spot candidates in the PCC consensus image surface. Background noise artifacts will heavily corrupt a gene spot that falls outside of this ring. The ring overlay is representative of the final VS determined gene spot shape and should be very similar to that determined by the human based GenePix identification process (see Section 2.4 for details). The ring was that as generated by the above discussed GHT gene spot location backup process designed to give an indication of the region's gene spot via a different (non PCC) approach.

FIGURE 6.13
VS layer one surface binding regions. Left: layer 1; Middle: layer 1; Right: layer 1

Generally, there is a slightly uneven spread of foreground and background pixels in the initial PCC layers. The GHT identification ring straddles the true gene spot regions fairly closely considering the noise and such present through the three surfaces. In fact, it appears that the GHT ring has achieved a better position in the region of the right image of Figure 6.13 (the worst gene spot of the set) than the clearer region of the middle image of Figure 6.13. The clarity of the right image is not really surprising given that the middle image surface has the large artifact section dissecting the gene spot itself, while the region of the right image is relatively stable. What is surprising for the right image surface is that the gene spot was found at all. Importantly, note also that the ET ring surrounds the gene spot adequately (and a major section of the GHT ring), across the three surfaces. In all cases, the SB process has identified good tentative gene spot locations as seen here and in Figure 6.12.

The VS process must now take the layer information and track the changes made to the region surfaces as the layer granularity parameter increases. These surface changes occur as the individual gene spot member pixels become ever

FIGURE 6.14

VS layer two surface binding regions. Left: layer 2; Middle: layer 2; Right: layer 2

more similar to the local background intensity as defined by the PCC grid-granularity parameter (see Section 4.3 for details). The images in Figure 6.14 represent the initial allegiance changes as the grid-granularity parameter of PCC transitions from grid size 2^2 up over. The remaining image surfaces of the gene spot regions will start to take advantage of the spatial aspects in PCC and result in related pixels coalescing into their appropriate groupings. Generally, the pixel spread slowly becomes biased to the gene spot areas themselves due to the spatial aspect.

Surface pixels in the 6.14 images start to change allegiances as a result of the local pixels in the PCC search-grid containing more or less foreground or background like intensities. These allegiance changing pixels specifically are what the VS process focuses on. The change in the allegiance is caused by a combination of grid-granularity parameter increases and similarity changes between the foreground and background domains within the new parameter defining space. Importantly, note that the result of the right image of Figure 6.14 contains more background element pixels than those of the other two. This bias is a direct result of the gene spot in the region of the right image of Figure 6.14 being negatively expressed against its local background (meaning there is more similarity between the gene spot and the background.

By the layer 3 images of Figure 6.15, the gene spot bias as mentioned above can be seen clearly. The pixels falling outside the ring region have started to be reclassified as background pixels. The pixels internal to the ring alternatively start to be reclassified as foreground if they were foreground at some point previously.

Again, the background regions of the images have been reduced into smaller distinct block areas. In the case of the middle image of Figure 6.15, the block areas present are those connected to the gene spot region. Although the interesting pixels do not lie in the background area, the VS process must examine these background pixels as the true determination can only be made once

FIGURE 6.15
VS layer three surface binding regions. Left: layer 3; Middle: layer 3; Right: layer 3

the VS step is complete. Importantly, note the foreground area of the region has also seen the reallocation of several gene spot pixels. The region of the right image of Figure 6.15 poses a bigger problem at this stage due to the foreground and background domains being very similar. Although the gene spot itself has been rebuilt slightly here, the background still contains a large number of contender pixels (and therefore gene spot member possibilities).

There are several layers present for the image set that see the reallocation of pixels taking place as the VS process refines the surfaces. Essentially, these surface reclassifications see the gene spot pixel contenders shifting their allegiance as the PCC search-grid changes. The majority of the background regions will have been reclassified as a consequence of this. Any remaining background blocks present in the regions now meet strict dimension criteria in order to remain in gene spot member contention. However these regions fail to meet the various ring criteria and are thus reclassified as background. These individual surface changes are not presented here for the sake of process clarity.

At the final stage of the process then, the VS component will have reclassified all gene spot pixel members. Ultimately, the pixels that are left in the image surfaces are those that fall within the various designator rings. All that now remains is to apply a surface consistency check to verify the match between the GHT, ET, and SB determined areas respectively. The match will rarely come back as perfect which means that the edges of the rings have to be compared to each other to generate the final gene spot surface areas themselves.

In the case of left image of Figure 6.16 this match process would discard the pixels in the top area of the ET region as these members are significantly different in intensity from the majority of the surface area. The same process is repeated for the remaining two image regions respectively. In the middle

FIGURE 6.16
VS final estimate surface binding regions. Left: final gene spot region; Middle: final gene spot region; Right: final gene spot region

image of Figure 6.16, several pixels are removed from the surface list. The right image of Figure 6.16 returns a similar process as the middle image where the green ring would have to be shifted slightly upwards to accommodate the PCC/ET positions.

6.3.3   Strengths and weaknesses

As per the IL and IS components, SB has performed well in the enforced blind environment. Generally, as long as the PCC component is able to take at least a partial glimpse of the gene spot in question, the SB process is able to expand upon this information and render a full gene spot profile. In situations where the gene spot was negatively expressed (the background intensity was higher than the gene spot), the SB process was still able to make a very good approximation of the gene spot location.

The VS process functionality was tied to a gene spot's region as determined by the IS component. From the above discussion this caveat has shown itself to be quite successful in the identification process. However, it is possible that casting the VS process back through the PCC time slices at the history layers granularity setting will help yield greater surface knowledge. A problem with this method would be the greater complexity generated for VS when tracking the individual pixel allegiances however.

Perhaps the greatest potential weakness of the SB process is associated with the initial gene spot tracking task. Here, a GHT process was used to determine the initial gene spot position within the region with the positioning process reinforced by the gene spot ET calculation step. Nevertheless, as the GHT method is specific to the individual gene spot (ET is based on all PCC known gene spots in the image) in situations where there is significant noise (especially false positives) in the region, the GHT method can produce a highly erroneous position designator. This designator can cause an increase in the SB process calculation as in essence, a gene spot's location and morphology will have to be calculated twice from completely different sets of pixels. Normally, the pixel sets from the designator rings will overlap somewhat, which means only those unique pixels from the sets have to be recalculated accordingly.

6.4 Evaluation of copasetic microarray analysis framework

With the completion of the SB process of the framework the gene spot pixels themselves can be quantified such that downstream analysis can be carried out. The final stage of the CMA framework therefore consists of components that can be thought of as the more conventional stages of microarray analysis. Using the structural information that has been previously identified along with the raw image data, the genes in the microarray images are summarized. This summary process consists of two stages, the first, Post Processing (such as background correction) and second, Final Analysis, a data reduction stage (such as converting the values from all the pixels that make up a gene spot into one representative result).

Background correction and normalization can take the gene locations that have been previously determined to perform local, global or combined correction techniques for noise removal (see for example O'Neill [157], Yang [226], Kepler [120], Geller [93], Kooperberg [123], and Quackenbush [168] for various techniques. Final analysis involves taking log_2 ratios of the two or more image channel gene spot templates as created by the earlier analysis stages. Such background and log transform approaches ensure that the noise and other unwanted pixel values will not be used in the quantification process.

With the gene spots from a microarray now quantified as based on the processes presented in the various chapters, a quality valuation can be made on the final CMA framework results. Importantly, note that such a validation process also indirectly quantifies the performance of the CMA framework's components as discussed. The validation process in essence takes the form of comparing the human optimized gene spot pixels (as determined with the GenePix analysis package) with those of the CMA framework's process results.

6.4.1 Peak signal-to-noise ratio for validation

In order to quantify the validation of the gene spot identification stage and therefore the performance capabilities of the CMA system as a whole, some form of quality measure is required. This measure will allow a comparison to be made between the GenePix and CMA processes. Both processes produce a template structure which classifies the underlying pixels as belonging to either the gene spot or the local background. By overlaying this template onto the original image, the metric can be used to quantify the differences that exist between the two pixel classification groups.

If the templates fit the gene spots closely, the separation between the gene spot and background groupings will be more distinct than if the opposite where true. There are many alternatives that could be used for this quality metric, a common one is the *mean squared error* (MSE) that is defined as follows:

$$\text{MSE} = \frac{1}{mn} \sum_{i}^{m} \sum_{j}^{m} \|I(i,j) - M(i,j)\|^2 \qquad (6.1)$$

where m and n are the image dimensions and I and M are the actual input and template images respectively.

These error metrics are computed on the luminance signal such that pixel values I fall in the range of zero (black) and one (white). For grey-scale images scaled between the values of zero and one, there are two disadvantages of the MSE percentage as defined in Equation (6.1). First, the denominator is usually very large when compared with the numerator. The difference between this denominator and numerator means that the reconstruction process improvement reduces this numerator value, but critically this change might not be observable. Second, the MSE metric is sensitive to the brightness of the original image. Therefore, a more objective image quality measurement known as the *peak signal-to-noise ratio* (PSNR) is used.

PSNR is used to define the ratio between the maximum observed value of a signal and the magnitude of that signal's background noise, with the PSNR result given in *decibel units* (dB). An increase of 20dB corresponds to a ten-fold decrease in the MSE difference between the two images. Therefore, the higher the PSNR value the more accurately the template fits with the original image's surface. The PSNR therefore gives an indication of the accuracy between the signals as defined by a microarray expert and the CMA techniques. The formula for PSNR is shown in Equation (6.2). PSNR is also more resilient to large ranges in an images intensity values unlike MSE [150].

$$\text{PSNR} = 20 \times \log_{10} \left(\frac{I_{\max}}{\sqrt{\text{MSE}}} \right) \qquad (6.2)$$

where $I_{\max}$ is the maximum pixel value in the image and MSE is the mean squared error.

A selection of results of the final PSNR comparison technique as applied to the GenePix and CMA process results are presented in Figure 6.17. In the figure, the GenePix and CMA values for template fitting are directly compared for individual microarray images.

The CMA results for the time series images (1-15) of the figure have not been directly improved over those of the GenePix process. However, the remaining

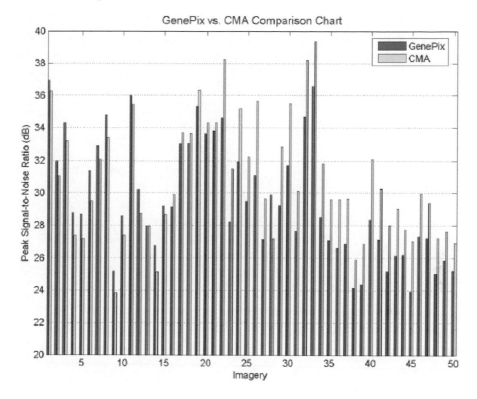

FIGURE 6.17

Final CMA result over test images

sample images show a marked 1-4dB improvement. Other than these initial time series images then, the CMA system has essentially outperformed the human expert in the task of gene spot identification when using GenePix.

6.4.2 Strengths and weaknesses

The application of the CMA framework process has been successfully applied to a range of microarray image datasets. Some of these images were considered high quality, this means their background regions were of low consistent levels, while the gene spots themselves were of higher intensity. Such differences between the two fundamental domains of an image saw the resultant gene spot identification process render high quality gene spot profiles. Unfortunately, microarrays typically produce images that are littered with high intensity background noise objects as was discussed in Section 3.2. These noisier microarray image surfaces present inherently more challenges during the subsequent analysis stages. Again though, the CMA framework process rendered results from these noisy images such that they were usually of higher quality than those as determined by the manual operator process (GenePix for example). Even in the situation where extremely noisy imagery is presented to the CMA framework, images that would typically be thrown away as being of no value are still rendered somewhat better than their manual counterparts.

The advantages associated with this fully automatic microarray analysis system are of considerable benefit to the laboratory researcher. Issues like time wastage (due to manual or semi-automatic addressing stages used in most literature processes) for the biologist. Experiment result repeatability (difficult to achieve with the above manual processes involved) and a consistent process flow, coalesce into a system that takes a step in the right direction for achieving the high-throughput potential of microarray hardware. The flexibility and adaptability built-in to the presented framework components have shown themselves to be more than capable of discovering an image's underlying structure and processing that structure appropriately.

For all of these advantages, however, the CMA framework is not perfect. From the test images used throughout the development process (most notably near the end of said process) it was seen that the framework does have some particular limitations. The most noticeable of these limitations is that of the negative gene spots issue. Specifically, if there are large groupings of negative gene spots present in an image, the IL and IS components of the framework will typically fail to correctly identify the gene block bounds within the dynamic (image specific consistency) criteria. In this case, this is one of the few areas in which the multiple redundancy of the framework's components have not corrected the slight errors in analysis made by a previous stage. The PCC result for such negative gene spot groupings indicates that such groupings will cause problems at later stages. However, the IL and IS stages use the results of previous components more directly. As such, the negative groupings can confuse the IL and IS algorithms by generating a partial shift of position within these regions. What should happen in this negative gene spot situation is that PCC should adapt the topology of the negative gene spots to be more similar to other domain gene spots (a reversal of intensity distribution if you will). As already discussed however, care should be taken with such a modification process as these negative gene spot positions are probable at this point, rather than confirmed. For example, an artifact of sufficient magnitude and morphology can render a similar situation as these negative gene spots, but who is to say which surface is the more valid. One could argue infact that it is the ITE rather than the PCC component that is the source of such errors. But again, care should be taken with modifications of process without further investigations into their knock on effects.

6.5 Summary

This chapter has investigated the performance of the *Spatial Binding* (SB) component for feature identification within the medical orientated images of microarrays. These images have small circular regions (gene spots) arranged across their surface into distinct block regions. Utilizing the results of the *Pyramidic Contextual Clustering* (PCC) and *Image Structure* (IS) components, the SB process is able to determine with a fair degree of accuracy and consistency a gene spot's location and morphology correctly.

The accuracy is achieved with the combination of a *generalized Hough transform* (GHT) and *estimate template* (ET) process working in conjunction with the aforementioned PCC framework results. The GHT provides a good starting location for the gene spot identification stage as it is highly robust to noise. The ET process reinforces this GHT location estimate by profiling the gene spots found in the entire PCC result. Once these provisional locations have been determined the *Vader Search* (VS) sub-process of SB examines each IS determined region as required.

Essentially, the VS process reclassifies a gene spot's pixel members by examining the individual pixel's allegiances through the PCC generated layers or "time slices." Due to the various sub-processes of the *Copasetic Microarray Analysis* (CMA) framework, this identification process is achieved with the smallest amount of human interaction possible as opposed to the GenePix type method. Once these gene spot locations and morphologies have been determined, the given gene's characteristics can be summarized via appropriate descriptive statistics.

The overall accuracy of the technique has been assessed utilizing a metric known as the *peak signal-to-noise ratio* (PSNR). This metric measures the disparity between two versions of an image. In this case these versions are defined by the gene spot regions as determined by the GenePix manual and CMA automated identification processes respectively. For the CMA framework to be deemed successful in its task, at a minimum, the framework would have to at least mimic the performance of the human operator on a set of images. In this regard, of the 122 microarray images used throughout the component verification processes, the CMA framework has performed well.

7

Feature Identification II

The spots printed onto a typical microarray slide are usually printed in separate subgrids each with the same number of rows and columns. The detection of these subgrids and the division of an image into its constituent subgrids is widely overlooked within the microarray image processing literature. The focus of most gridding papers is upon the task of detecting the rows and columns of spots within a single subgrid. Many papers on the subject of gridding either do not mention the task of subgrid detection or state that it can be handed manually. In Jung and Cho [114], the observation has been made that approaches that do not include automated subgrid/block detection can hardly be described as a fully automated processing system. The approaches highlighted in Jung and Cho [114] are ScanAlyze [184], GenePix [14], AutoGene [28] and the approach from [199]. Indeed, subgrid detection is an important task in the processing of an image and therefore is required in a complete framework. Possibly the reason for this problem being widely overlooked is because the task of mapping the subgrids can at first appear trivial and suitable as a manual task when compared to the task of subgrid addressing. An image will typically contain thousands of spots but only a few dozen subgrids. However, since the creation of a totally automated framework is desirable, no task should be treated more trivially than any other and completely blind subgrid detection presents a unique set of challenges. In Bozinov [33] the task of subgrid detection (or metagrid detection) is described as "one of the hardest challenges."

As with spot detection, within this chapter the task of subgrid detection will be discussed, with all issues that make this task difficult highlighted. Previous work in this area will then be discussed with weaknesses of past approaches identified. A new approach to this problem developed as part of this book is then described and tested. The results of testing this new approach show that it is an improvement upon all previous techniques and even offers applications within the broader topic of image analysis.

7.1 Background

A microarray image contains one or more regularly spaced grids of spots (given a small degree of printing error). Each of these grids is referred to within

FIGURE 7.1

Two example microarray images with differing subgrid arrangements

the microarray processing literature as a subgrid. Each subgrid contains the same number of rows and columns of spots (although the number of visible spots will certainly differ from subgrid to subgrid). A typical image could, for example, contain 24 subgrids of spots arranged into 2 columns each containing 12 subgrids, each subgrid containing 32 columns and 12 rows of spots. Such an image is shown within Figure 7.1 (the image on the left). Figure 7.1 also shows an image with a different structure (the image on the right) with 48 subgrids arranged into 4 columns.

The task of locating the subgrids within an image is complicated by all of the noise sources. In addition, the task can be complicated further due to 1) the fact that there is often a small amount of drift between subgrids; 2) the fact that entire rows and/or columns of spots may not be visible because of low levels of expression, which results in the appearance that some subgrids are smaller than others; and 3) the fact that spots within some subgrids may be more tightly packed than in others because of errors in the printing process, giving rise to subgrids that actually are smaller than others (cover a slightly smaller image area).

Automated subgrid division is a much overlooked area within the microarray processing literature. Although a great deal of time has been spent researching automated microarray gridding, the focus has tended to be away from the task of subgrid detection. Many gridding papers in the literature do not even offer a solution for this critical phase of microarray image processing instead focusing upon the subsequent task of subgrid addressing (identifying the rows and columns within an individual subgrid). A possible explanation for this could be that the proposed approaches are intended for use on images featuring only one grid of spots (this is the case in Steinbath et al. [199] and Brandle et al. [36]).

Many previous approaches for subgrid detection require several input parameters from the user, such as the number of subgrids within the image, and the arrangement of the subgrids (number of rows, number of columns). For example, see Jain et al. [111] where the user is required to enter the geometry of the subgrids as well as the number of rows and columns of spots that each contains. Clearly a framework that can automatically process images featuring an unspecified number of subgrids in an unspecified arrangement is of more use than one that cannot. A framework that takes only the raw image as input is clearly more powerful than one that requires additional input from the user. Also, by taking parameters such as the number of rows and columns within an image as input, it is inherently assumed that the subgrids within the image are arranged in a regular 2D grid pattern (x rows by y columns). While this is a relatively safe assumption to make with the current generation of microarray technology, future changes in the technology may well result in changes to the structure of the images and it is therefore of benefit to have a subgrid detection algorithm that is robust against such changes.

Microarray gridding (or subgridding) approaches are generally viewed as either being template-based or data-driven with various levels of automation [17]. Template based approaches can only be applied to images that do not feature any significant level of deviation from the expected model so are not suitable for widespread use [14]. The vast majority of data-driven approaches are based around the use of 1D projections, so much so that the term "data-driven" has almost become synonymous with "1D projection" in regard to microarray image processing. 1D projections are typically created by summing an image along its horizontal and vertical axes. The summation may not necessarily be of the raw image values but can be of some transform of the image, for instance the output of an edge detector [16]. For more examples of the widespread use of 1D Projection analysis in microarray image processing see [16,30,33,36,76,97,103,111,130,135,199,211]. Unfortunately, there appear to be several problems when the 1D projection algorithms are applied to microarray image processing.

The use of basic 1D projections is based around the assumption that a microarray image can be divided into its constituent subgrids with straight lines drawn from one side of the image to another. While currently the most commonly found microarray technology does print subgrids in a 2D structure it is conceivable that as microarray technology evolves the structure of the images will change, so it is desirable to have a subgrid detection algorithm, which is future proof, in the sense that it can detect subgrids of spots whether they are arranged in a 2D structure or in any other arrangement.

On the other hand, in the case that the subgrids within an image are indeed printed in a 2D pattern, there are still two issues that can severely complicate the use of 1D projections: 1) The first issue is that microarray images can feature varying degrees of rotation, since subgrid edges are not always parallel to the image edges. This problem can be countered by calculating the angle of rotation and either rotating the image back through this angle or using the calculated angle value as a parameter in subsequent gridding steps [16, 33, 36, 103, 199, 211]. Unfortunately, many of the methods used to

FIGURE 7.2

An image featuring significant subgrid "drift"

calculate the rotation angle require multiple rotations of the image, due to the typical dimensions of a microarray image, and are therefore computationally expensive. Additionally, rotating the image back to solve the problem is undesirable as it will alter spot morphology within the image. 2) As stated earlier, some microarray printing devices are capable of misprinting subgrids away from their desired locations, so it is not always possible to successfully segment an image using only straight lines. Figures 7.2 and 7.3 illustrate this problem. Figure 7.2 shows a microarray image (included in the test set for the new approach described later in this chapter) which features a clearly visible subgrid "drift." Figure 7.3 shows the same image with a vertical line placed alongside the left edge of the top right subgrid, and this line intersects with the bottom left subgrid, illustrating that with this image subgrid separation is not possible using straight vertical and horizontal lines.

Motivated by the above discussion, one of the goals within this PhD project is to develop a method of subgrid detection that does not rely upon 1D projections. An additional goal is to develop a method that does not make any assumptions regarding the shape, number or arrangement of the subgrids within an image, in other words, completely blind microarray subgrid detec-

FIGURE 7.3

The same image from Figure 7.2, with a vertical line illustrating the difficulty of dividing the image into its component subgrids using straight lines

tion. The addressed task can be more formally defined as locating regularly occurring groupings of a primitive (spot) within an image, where each of the groupings contains approximately the same arrangement of primitives. In the next sections of this chapter a method to accomplish this task will be proposed and evaluated.

7.2   Proposed approach – subgrid detection

It has been shown that, when filtering an image for spots it is advantageous to first determine as much information about the size and shape of the spots within the image as possible. Similarly, when dealing with the task of locating subgrids within an image, it logically follows that one should begin by determining some descriptive information about the subgrids prior to searching for them within the image. Therefore, as part of this book, a method to detect the approximate shape and dimensions of the subgrids within an image will be developed, and the information obtained will then be used to search the image for its subgrids. These sub-steps will be included in a new approach designed to accomplish the task of blind microarray subgrid detection.

In this section the main steps of the new approach to detect and divide an image into its constituent subgrids are described and evaluated. The main steps are 1) filter the image, 2) calculate the approximate spot spacing distance (the distance between two neighboring spots in the same subgrid), 3) detect the approximate shape of the subgrids, and 4) detect all of the subgrids within the image. These steps are shown in Figure 7.4.

Each of the main steps shown above will now be discussed in turn, and where appropriate individual evaluations will be made. After each of the steps have been discussed the entire approach will be evaluated by applying it to a series of test images including 50 real microarray images and a series of artificial test images.

7.2.1   Step 1: Filter the image

Microarray images can feature high levels of noise and high percentages of poorly expressed spots. It is therefore desirable to filter the image to remove any noise signals. It would also be desirable to attempt to strengthen the signal of locations that resemble the target of the search, in this case the subgrids. However, at this stage, the only information that can be assumed about the subgrids is that they are comprised of groupings of a known primitive (spot) shape, where the primitive shape can be considered known as it can be calculated using the new technique proposed in section spot shape detection, and the dimensions of the groupings themselves is still unknown. Therefore,

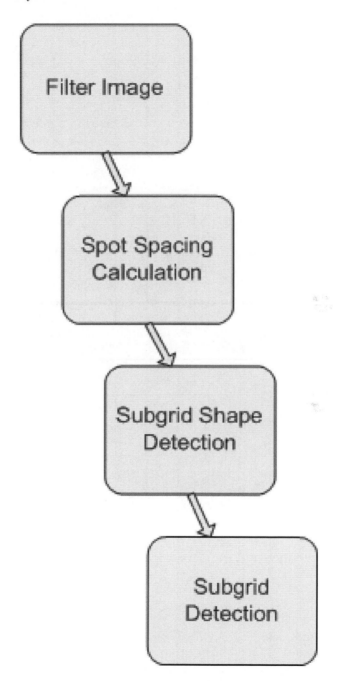

FIGURE 7.4

The main steps in the proposed method for subgrid detection

it is logical to amplify the signal of the spot locations. The filtering approach is ideal for this task, returning a real valued image where the highest value pixels should indicate the locations of spot centers.

7.2.2　Step 2: Spot spacing calculation

To scan an image for subgrids as thoroughly as possible it is first necessary to identify the size and shape of the subgrids within the image. In the next section a method for the detection of the shape of the subgrids within an image will be proposed. This new method requires three inputs, the spot filtered image (SFI), the approximate horizontal and vertical distances between adjacent spots within the same subgrid. As part of this book, a method for calculating these values from a spot filtered image is developed, which is based on the following three assumptions: 1) In the SFI of an image, the majority of high valued pixels (say pixels with a value of 0.7 or higher) are indicators of spot centers. This assumption is justified as the high valued pixels are the function of the spot matching filter. As was demonstrated previously, the filter works very well at highlighting spot center pixels. 2) The majority of spots within the image can be detected by the filter so that, for the average detected spot, one or more of its neighboring spots can also be present. This assumption is also justified by way of the 1D projections. Clearly the filter succeeds in highlighting the majority of the spots within an image. 3) Rotation of the image cannot occur past 45 degrees. This is a safe assumption to make with the current generation of microarray technology as rotation does not usually occur by more than ±5 degrees.

The approach for spot spacing calculation can be broken into three sub-steps: 1) The SFI is thresholded with all high valued pixels assumed to indicate spot locations (for groups of connected pixels the center pixel is taken as the spot center). 2) For each detected spot location the distance to the closest spot is within ±45 degrees of the horizontal and vertical. The distances to these two spots are added to 2 lists: horizontalSpacing and verticalSpacing. 3) The median of each of these lists are taken as the approximate spot spacing distances for the image and stored in two variables hSpacing and vSpacing.

The first sub-step of the approach serves to create a list of spot locations. Subjecting the output of the filtering process to a relatively high valued threshold (say 0.7), a large number of spots can be identified from within an image with few false positives. Assuming that this set of spot locations contains a large number of neighboring spots (spots from adjacent rows/columns), it is clear that the median of the list of distances between each spot and its closest neighbor in the horizontal plane will reflect the column spacing value. Correspondingly, working in the vertical plane should return a value that reflects the row spacing value.

The accuracy of this process is not critical as long as it is a "ball park figure," and the values will be sufficient for the subsequent stages of the subgrid detection process described throughout the rest of this chapter. Verification of this proposed approach can be inferred from the results of the spot shape detection method and the results of the subgrid detection method, both of which are proposed in the two following sections. If the calculated spacing values are not suitable then clearly neither of these processes will function correctly.

7.2.3   Step 3: Subgrid shape detection

In this part, a method of subgrid shape detection is documented. The only inputs to this process are a real valued spot filtered image (SFI) (created from a raw microarray image using the filtering approach) and the horizontal and vertical spot spacing distances (in pixels) (hSpacing and vSpacing). The key idea behind this process is that, if working with a perfect image subjecting the SFI created from the image to a threshold and then dilating the result with a structuring element with dimensions approximately equal to the spot spacing values, the resulting binary image should feature connected components that identify each of the subgrids within the image. A perfect image in this context can be thought of as an image with no noise and all spots regularly shaped and expressed above the background level of the image, where the distance between spots in the same subgrid is smaller than the distance between any two spots in different subgrids.

Basic morphological functions such as dilation, erosion, opening and closing have been used to process microarray images in the past. In [11], morphological operations have been used to filter out aspects of the image. These operations have been performed on a grey scale version of the image, as opposed to the black and white image created by the threshold operation as proposed here.

Figure 7.5 shows a manually created "perfect image" with 4 subgrids of 5 rows and 5 columns and no noise. Figure 7.6 shows the result of dilating the image with a rectangular structuring element with dimensions equal to the calculated hSpacing and vSpacing values for this image.

In reality it is not possible with most microarray images to perform a single threshold and dilation upon the SFI to identify every subgrid. This is mainly due to high numbers of poorly expressed spots which will disappear after the threshold. In addition, if the threshold is set too low, then false positives between subgrids will merge them together during the dilation operation. The process therefore uses two internal variables: `thresh` and `dilateMultiplier`. Initial values for `thresh` and `dilateMultiplier` can be randomly set, with `thresh` taking any value between 0 and 1 (the full range of values of similarity in the SFI) and `dilateMultiplier` taking any value between 0.5 and 2. These variables allow for a series of binary images to be created, with the hope being that in at least one of the images several of the images' subgrids are highlighted.

The reason that the internal variable `dilationMultiplier` can take any value between 0.5 and 2 is given as follows. If the calculated spacing value is slightly inaccurate or if some spots in the image have a slight variance in

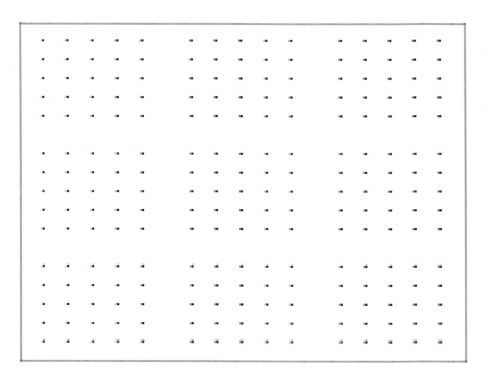

FIGURE 7.5
An artificial "perfect image," featuring 9 subgrids

FIGURE 7.6

Figure 7.5 dilated with a rectangular structuring element

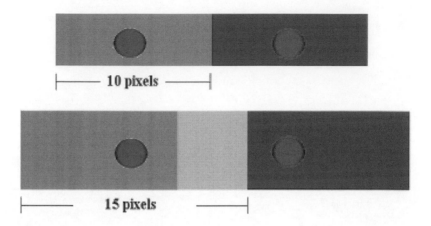

FIGURE 7.7

Illustration of the dilation of two spot centers with a structuring element that results in one connected mass

their spacing, then using a single structuring element with the dimensions of the calculated spacing variables will probably produce undesirable results. This will occur even if all spot centers have been detected with no false positives. If the spacing values are too large, then neighboring subgrids may be dilated together, too small and a subgrid's spots will be disconnected after the dilation. For example, if two spots are horizontally 10 pixels apart, then dilating both of the spot centers with a structuring element of width 10 will result in two boxes that only just touch one another in the output. Figure 7.7 illustrates this example. The left area shows the area covered by the dilation of the left spot and the right area, the area from the right spot. In the top half of the figure, the dilation is performed with a structuring element of width 10 pixels, where the two dilations only just touch. In the bottom half, a width of 15 pixels (1.5× the horizontal spacing distance) is used, where there is an overlap (the middle area) strongly connecting the two regions together.

By allowing `dilationMultiplier` to take values smaller than 1, it will counter for any overcalculation of the spacing variable values. Spot drift or undercalculation of the spacing variable values is countered by allowing `dilationMultiplier` to take values greater than 1. Bounding the variable to the range of 0.5 to 2 reduces the search space and has proven experimentally to contain a suitable value for all images tested to date.

While it is not possible to always identify every subgrid with a single threshold and dilation, it is typically possible to threshold and dilate a microarray image's SFI so that many subgrids are successfully identified. Figure 7.8 shows

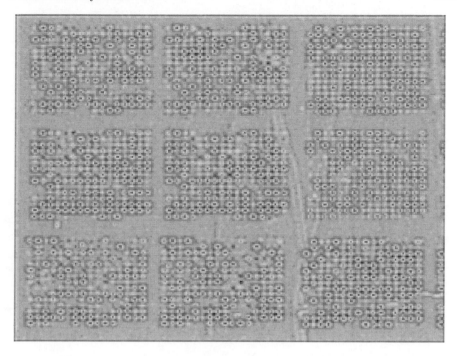

FIGURE 7.8
An area from the SFI of a typical microarray image

an area from a typical image's SFI. This image is subjected to a binary threshold (**thresh** = 0.5) and dilated using a rectangular structuring element, the dimensions of which are calculated by multiplying the hSpacing and vSpacing values by the **dilationMultiplier** (in this case **dilationMultiplier** is set to 1.5), and the resulting image is shown in Figure 7.9. It is seen clearly in Figure 7.9 that nine large connected components covering the majority of the nine subgrids from the original image have been formed by the dilation.

To obtain the approximate shape of the subgrids within the image, the mass of each object within the image is calculated and any small objects (with a mass smaller than say 25% of the largest object) are disregarded, which will remove the small objects usually created by isolated spots (spots whose neighbors are removed during the threshold operation) or sometimes false positives. The remaining objects are compared to each other. Note that there are many methods of calculating a similarity value between objects, and the decided measure is arbitrarily chosen as height and width. The object that is found to closely resemble the most objects in the image and all of the objects that closely resemble it are then merged together to form an estimate of the subgrid shape. Figure 7.10 shows the output of this process on the image in Figure 7.8.

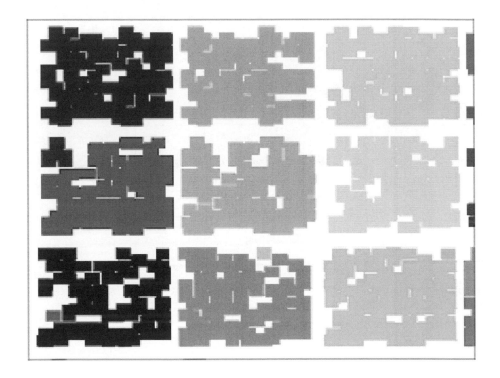

FIGURE 7.9
Same area from Figure 7.8 having been subjected a threshold and a dilation.
Each connected object is displayed in a different color/shade

FIGURE 7.10
The approximate subgrid shape

A reliability measure for the output shape can be calculated by dividing the mass of all objects that resemble the representative object by the total mass of the image after the initial dilation. If the reliability measure is below a predefined value (set to 0.5 in the following code), then the process can be repeated with a different value for `thresh` and/or a different value for the internal variable `dilationMultiplier`. If the reliability measure does not meet the predefined value after several iterations (set to 10 in the below code), then the shape with the highest reliability score can be taken as the subgrid shape. The detected shape is stored into the variable `subGridShape`.

The outline of the whole subgrid shape detection process is as follows:

```
1  Input: SFI, hSpacing, vSpacing
2  bestScore = 0
3  bestShape = null
4  rScore = 0
5  loopCounter = 0
6  while rScore < 0.5 AND loopCounter < 10
7      loopCounter = loopCounter + 1
8      dilateMultiplier = random value between 0.5 and 2
9      thresh = random value between 0 and 1
10
11     Create a rectangular structuring element se with
           dimensions of hSpacing*dilateMultiplier by ySpacing*
           dilateMultiplier.
12
13     temp = (SFI > thresh) + (SFI < thresh*-1)
14     dilate temp with se
15     mass1 = sum(Temp)
16
17     find the mass of the largest connected object in temp
           remove any objects that are smaller than 25% of this
           value
18
19     tempShape = the most frequently occurring shape in temp
20     mass2 = sum the mass of all objects in temp that resemble
           tempShape
21     rScore = mass2/mass1
22     if rScore > bestScore then
23         bestScore = rScore
24         bestShape = merge all shapes in temp that resemble
               tempShape
25     end if
26 end while
27
28 output: rScore, bestShape
```

Algorithm 7.1: Subgrid Shape Function

With the subgrid shape determined, the next step is to detect all of the subgrids within the image, but before a method to accomplish this is proposed, results of experiments with the subgrid shape detection process will be briefly discussed.

7.2.3.1   Experimental results – subgrid shape detection

The new subgrid shape detection process is at first tested with a series of artificial images. Each of the images features several regular groupings of a primitive, and the primitives within each grouping are closer to each other than to any primitive in any other grouping. The process is also tested on a series of 50 microarray images. In all images the process successfully returns a shape that is a very good (results analyzed manually) representative for the subgrids in the image.

These results will be presented in greater detail as a part of the final results of the subgrid detection process towards the end of this chapter.

7.2.4   Step 4: SubGrid detection

With the subgrid shape detected, the next logical step is to locate all of the subgrids within the image. At an initial glance this can appear to be a similar problem to that of spot identification, but the task here is in fact slightly more complicated. The subgrids being composed of many unconnected components (the spots) mean that a simple comparison of each connected object within the image to the calculated subgrid mask cannot be performed (unlike the spot detection process). Instead, the objects (spots) within the image need to be grouped together in some fashion and their groupings then compared to the subgrid mask.

Using the SFI, `subGridShape` and the row and column spot spacing values (hSpacing and vSpacing), it is possible to search the image for subgrids. As was illustrated in Step 3, subjecting the SFI to a binary threshold then dilating the result will often identify many subgrids within the image. However, it is typically not possible to identify a single pair of values for `thresh` and `dilationMultiplier` that will successfully identify every individual subgrid in the image.

In view of the above discussion, a threshold and dilation based approach used to locate all subgrids within the image must take multiple views of the image created with a range of values for thresh and `dilationMultiplier`. The set of threshold values used in the experiment to test this approach are between 0.3 and 1 (values below 0.3 are manually determined to contain a considerable number of false positives), and the parameter `dilationMultiplier` is set to take values between 0.5 and 2. The notion that multiple views of a microarray image are often required to obtain better knowledge of the image was previously exploited in [84] although not specifically for the task of subgrid detection.

After each threshold and dilation operation the output image is searched for subgrids. This is accomplished by comparing each of the connected objects within the new image to the known subgrid shape. Any objects that closely resemble `subGridShape` are added to the final output image finalOutput provided that they do not significantly overlap any other objects which have previously been added to finalOutput.

Figures 7.11–7.14 show the state of the finalOutput over 4 iterations when the algorithm is used on one of the microarray images from Figure 7.1 (the image on the left). The initial processing of the image is performed with `thresh` = 0.3 and `dilateMultiplier` = 2 (see Figure 7.11), which correctly identifies 16 of the 24 subgrids within the image. Figure 7.12 shows the second iteration with `thresh` = 0.3 and `dilateMultiplier` = 1.5, where two more subgrids are added to the finalOutput. Figure 7.13 shows the finalOutput after `thresh` = 0.3 and `dilateMultiplier` = 1, with 4 more subgrids identified. In Figure 7.14, with `thresh` = 0.5 and `dilateMultiplier` = 2, the remaining two subgrids are added to the finalOutput.

The algorithmic description of the process is given in Algorithm 7.2.

```
 1  Input: SFI, subGridShape, hSpacing, vSpacing
 2  finalOutput = emptyimage, the same size as SFI
 3  terminate = false
 4  thresh = 0.3
 5  while thresh < 1 AND terminate = false
 6      dilateMultiplier = 0.5
 7      while dilateMultiplier <= 2 AND terminate = false
 8          Create a rectangular structuring element se with
                  dimensions of hSpacing*dilateMultiplier by ySpacing
                  *dilateMultiplier.
 9
10          temp = (SFI > thresh) + (SFI < thresh*-1) dilate temp
                  with se
11
12          find any shapes in temp that resemble subGridShape
13
14          add shapes to outputImage, provided they don't overlap
                  any shapes already in the image
15
16          terminate = imageFull (finalOutput, subGridShape)
17      end while
18      thresh = thresh + 0.1
19  end while
20  Output: finalOutput
```

Algorithm 7.2: Locate Subgrids Function

A time saving device has been built into the algorithm, that is, if the output image can no longer fit any more subgrid shapes (without overlapping shapes that are already in the image) then the process will terminate. This can be accomplished by eroding an inverse of the binary output image with a structuring element that is the same shape as the `subGridShape`. The process is in the function imageFull and is described as follows:

```
 1  Input: finalOutput, se
 2      temp = finalOutput > 0
 3      temp = temp * -1 + 1
 4      test = erode temp by se
 5      if sum(test) == 0 then
 6          x = true
```

FIGURE 7.11
thresh $= 0.3$ and dilateMultiplier $= 2$

FIGURE 7.12
thresh = 0.3 and dilateMultiplier = 1.5

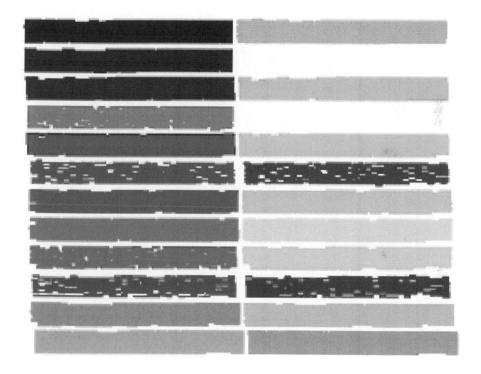

FIGURE 7.13
thresh $= 0.3$ and dilateMultiplier $= 1$

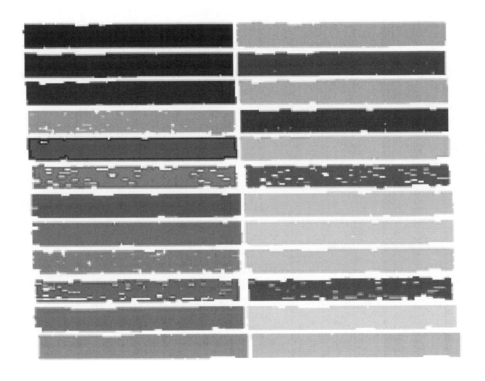

FIGURE 7.14
thresh $= 0.5$ and dilateMultiplier $= 2$

```
 7      else
 8            x = false
 9 Output: x
10 End Function
```

Algorithm 7.3: Image Full Test Function

Once the process has terminated it is beneficial to subsequent stages to place the known subGridShape over each of the detected subgrids. The most obvious benefit is that many of the detected subgrids will contain gaps and holes that will be filled by placing the subGridShape on top. Figure 7.15 shows an example of the finalOutput from a typical image with many of the subgrids featuring small holes and chips around the edges. The 6th subgrid down in the right hand column features a very noticeable missing region. Figure 7.16 shows the same image after the subGridShape is added on top of each subgrid. The improvement is clear.

Some subgrids may be slightly smaller in either width or height than the known subGridShape due to entire rows/columns of poorly expressed spots. When the known subGridShape is wider than a detected subgrid the output image is tested to see whether the placement of the known subGridShape against the detected subgrid's right or left edge will overlap any other grids. If the placement is possible without any overlap then it is made. The same operation is performed with respect to the subgrid's height and the top and bottom edges. Figure 7.17 shows the finalOutput from another image in the test set, and this image includes several subgrids which have several missing columns on their right hand sides (bottom right corner). Figure 7.18 shows the finalOutput after the edge placement process is performed.

It is desirable to assign every pixel in the output image to a specific subgrid, which means that the output image can then be used as a subgrid map for the original microarray image. This can be accomplished by dilating every subgrid iteratively by a single pixel using a 3×3 structuring element, with any subgrid pixels created by the dilation which overlap any other subgrid are removed. Figure 7.19 shows the result of these dilations on the image from Figure 7.18.

7.3 Experimental results

The most logical, and also the simplest way of testing the approach is to apply it to a set of real microarray images. The only input given to the algorithm for each image is the image itself and no data is carried over between images. The proposed approach is tested with a set of 50 images, comprising two types of image. The first features 24 subgrids arranged in a 12×2 format (see left hand side of Figure 7.1), and the second contains 48 subgrids arranged in

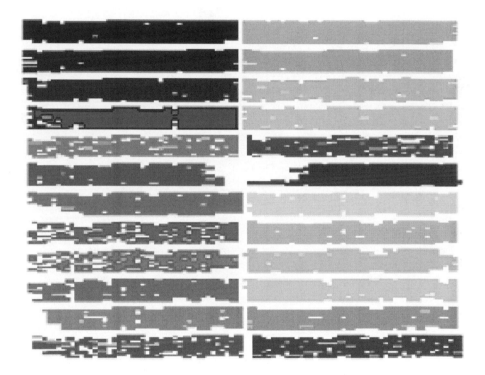

FIGURE 7.15
finalOutput image featuring many holes and one subgrid with a very noticeable chunk missing

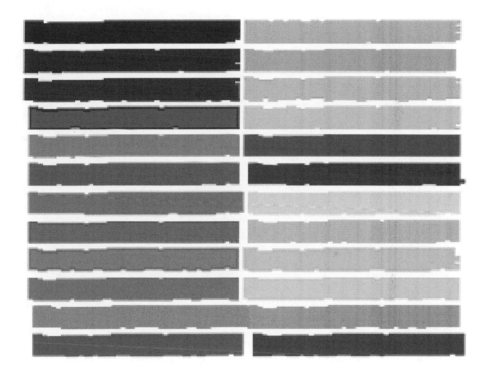

FIGURE 7.16
The image from Figure 7.15 having had the `subGridShape` placed over each subgrid

FIGURE 7.17
finalOutput from a typical image featuring several subgrids with missing columns

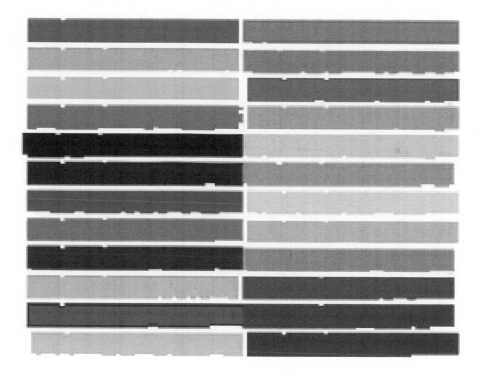

FIGURE 7.18

Figure 7.17 after edge placement process

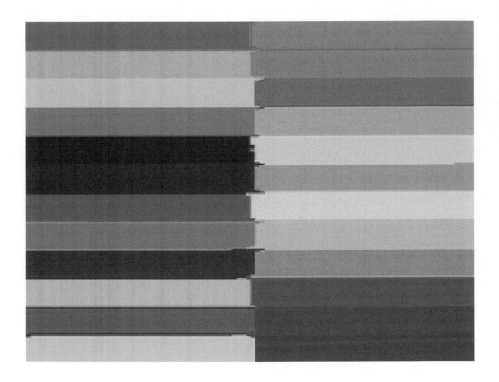

FIGURE 7.19
The result of applying the subgrid dilation process to Figure 7.18

12×4 structure (see right hand side of Figure 7.1). The quality of the images is typical of the varying quality of images seen in microarray technology; some relatively clean, others with very high levels of noise and/or high numbers of poorly expressed spots. In all 50 images the algorithm has successfully divided the images into their constituent subgrids, with all spots correctly contained in the appropriate grid.

Some of the poorest quality images are shown in this section to illustrate the robustness of this approach. Figure 7.20 shows a microarray image which features some very sparse subgrids. Each subgrid in the image contains 384 spots (12 rows and 32 columns). In the top left subgrid only 22 spots are visible - less than 6%. The top left subgrid is shown close up in Figure 7.22. One of the lower subgrids is shown in Figure 7.23 for contrast. The approach is still able to successfully detect all 24 subgrids within this image. Figure 7.21 shows the grid map for the image.

Figure 7.24 shows two images with very high levels of noise as well as their corresponding grid maps created using the approach. All subgrids are successfully identified, and valid boundaries created. It should be noted that, in the top left image featured in Figure 7.24 (and several other images in the test set), the level of background noise does severely hinder/prevent correct gridding using some of the traditional 1D approaches.

After testing with all 50 images at their original size, the tests were repeated with resized versions of all 50 images at 50% and 25% of the original size. The results are identical, in all images the subgrids are correctly identified, with all spots being contained within the correct region. As well as demonstrating the robustness of the approach this also provides a run time saving mechanism. By working with smaller images the run time is significantly reduced. Figure 7.25 shows the run time of the entire process (hSpacing and vSpacing calculation, subgrid shape detection and subgrid detection) for a typical image at 4 different sizes.

The algorithm is also tested with a series of artificial images, aiming to test its robustness against differing primitive shapes and differing arrangements of subgrid primitives as well as subgrid arrangements. This also contributes towards testing of the spot shape detection and filtering approaches. Figure 7.26 shows one of the artificial test images, which is an image with regular groupings of triangular primitives and the groupings arranged randomly. Figure 7.26 also shows the output of the approach, indicating that the algorithm successfully determines the shape of the primitive, the shape of the groupings of the primitives within the image, and then identifies all of the groupings in the image.

Figure 7.27 shows another artificial image from the test set. This image features triangular groupings of star shaped primitives. The groupings are arranged in an "orange packing" structure. Figure 7.27 also shows the output of the approach with all groupings successfully identified.

As we discussed before, the more traditional 1D based approaches are unable to successfully demarcate the artificial images featured here, illustrating

FIGURE 7.20
An image featuring several subgrids with very few visible spots

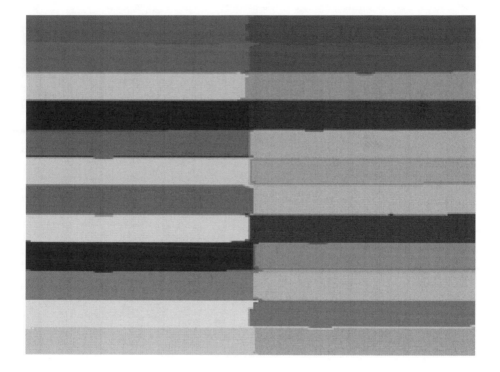

FIGURE 7.21
The grid map for the image from Figure 7.20

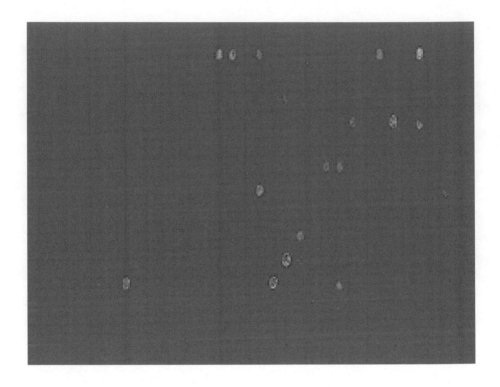

FIGURE 7.22
The top left subgrid from the image shown in Figure 7.20

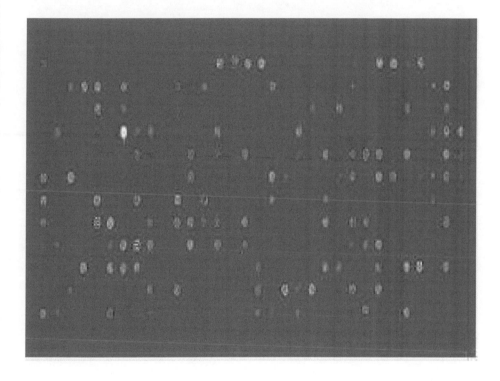

FIGURE 7.23

A more visible subgrid from the image shown in Figure 7.20

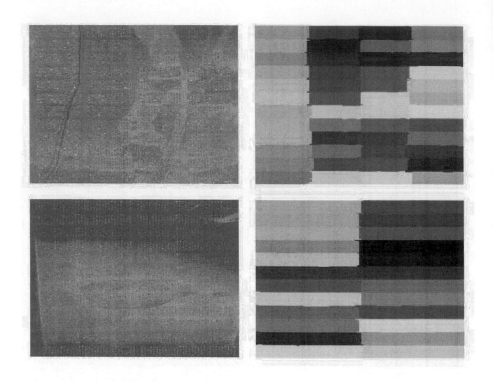

FIGURE 7.24
Two particularly noisy images and their subgrid maps

Size Scale (%)	Run time (seconds)
100	1223
50	94
25	47
16.666	25

FIGURE 7.25
Subgrid detection process run times at various resized scales

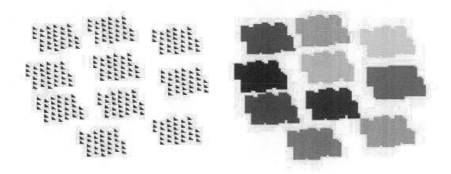

FIGURE 7.26
An image from the artificial test set and its subgrid map

FIGURE 7.27
Another image from the artificial test set and its subgrid map

that the proposed approach is more likely to cope with any changes in microarray technology and also demonstrating that the algorithms may have applications outside of the domain of microarray image processing in the broader area of image processing.

7.4 Conclusions

In this chapter, it has been proven that the developed approach for microarray subgrid detection is robust against high levels of noise, high percentages of missing spots and all of the other factors that complicate the task of microarray subgrid detection. The algorithm(s) tested have also proven to be potentially robust against future changes in microarray technology as they can cope with different shaped primitives, different subgrid configurations and subgrid distributions. The approach can also work with a range of image resolutions, offering time saving benefits and proving robustness against an increase/decrease in primitive size and scanner resolution. The proposed method is clearly more robust than the 1D projection based approaches and is more likely to be able to cope with any new development in microarray technology. The method may also offer application in other areas, as proven with the artificial image set, providing a method of identifying and locating regularly appearing groupings of the same high valued primitive within an image.

8

Chained Fourier Background Reconstruction

8.1 Introduction

With the invention of microarray technology in the mid-1990s, modern day genetics research underwent a paradigm shift. These microarray devices facilitate the real-time monitoring of tens of thousands of gene expressions simultaneously. With a so-called "gene chip" containing probes for an organism's entire transcriptome, differing cell lines thus render gene lists with appropriate activation levels. The gene lists can be analyzed with the application of various computational techniques, for example clustering [70], or modelling [119] such that differential expressions are translated into a better understanding of the underlying biological phenomena.

As with any real-world data analysis problem however, addressing the issue of data quality effectively is a major challenge. No matter how high development values are set; space will always be available for cracks to appear. In this context the cracks surface as "noise" within an output slide image. This noise can take many forms, ranging from common artifacts such as hair, dust and scratches on the slide, to technical errors like the random variation in scanning laser intensity or the miscalculation of gene expression due to alignment issues. Alongside these technical errors, there exists a host of biological related artifacts such as contamination of complementary Deoxyribonucleic Acid (cDNA) solution or inconsistent hybridization of the multiple samples. Much work in the microarray field therefore focuses on analyzing the gene expression ratios themselves [44, 70, 92, 120, 167, 168] as rendered from the image sets. This means there is relatively little work directed at improving the original images to begin with References [157, 227] such that the final expressions are more realistic.

Noise generated in the images (even in such highly constrained environments) can have a detrimental effect on the correct identification and quantification of the underlying genes. Therefore, in this chapter, we present an algorithm that attempts to remove the biological experiment from the image. Recall, a microarray image essentially contains two layers; the first (bottom) layer consists of the glass substrate material upon which the gene spots will be placed, while the second (top) layer holds the gene spots. Ideally, this bottom layer will be optically flat, i.e., the entire surface will consist of one value.

Realistically however, this is very difficult to achieve in a reliable fashion. Which means this bottom layer consists of ever changing signals, although such signals are usually low-level. In the microarray field, it is accepted as part of the methodology that this background domain infringes on the gene spot's valid measure and steps must therefore be taken to remove said background from the foreground during quantification. The purpose of the removal process therefore is to "clear" the top layer such that the hidden regions of the bottom layer become visible. In effect, this removal process is equivalent to background reconstruction and should therefore produce an image which resembles the "ideal" background more closely in the experimental regions. Subtracting this new background image from the original image would in-turn yield more accurate gene spot regions. Gene expressions rendered by this reconstruction process are contrasted to those as produced by GenePix [14] (a commercial system commonly used by biologists to analyze images) and O'Neill et al. [157] (one of the first reconstruction processes implemented to deal with microarray image data specifically).

The chapter is organised in the following manner. First, we formalise the problem area as it pertains to microarray image data and briefly explain the workings of contemporary approaches in Section 8.2. Section 8.3 discusses the fundamental idea of our approach with the appropriate steps involved in the analysis highlighted. Section 8.4 then briefly describes the data used throughout the work and evaluates the tests carried out over both synthetic and real-world data. Our findings are then summarized in Section 8.5 with some future directions provided.

8.2   Existing techniques

As microarray image analysis techniques require knowledge of a given gene's approximate central pixel and the slide's structural layout, they all have similarities (regardless of their specific implementations). For example, a boundary is defined around the gene - thus marking the foreground region - with any outer pixels in a given radius taken to be local background. The background median is subtracted from the foreground and the result is summarized as a $\log_2$ ratio. Other bounding mechanisms include pixel partitioning via histogram [44,169] and region growing [1,9] functions with a detailed comparison of the more common approaches given in [227]. The underlying assumption for these mechanisms is that there is little variation within the gene and background regions.

Unfortunately, this is not always the case as seen in Figures 8.1-8.2, which depicts a typical test set slide (enhanced to show gene spot locations) with a total of 9216 gene regions on the surface and measuring $\sim$5000$\times$2000 pixels.

A good example of the low-level signal produced in the image can be seen in the close-up sections, where problems such as missing or partial gene spots, shape inconsistencies, and background variation are evident. Such issues are further highlighted in b and c where the scratch and background illuminations around the presented genes change significantly.

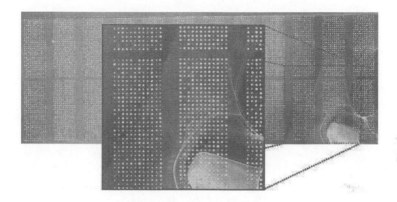

FIGURE 8.1
Example images: typical test set slide illustrating structure and noise

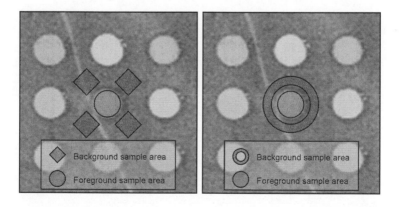

FIGURE 8.2
Sample gene, background locations for GenePix valleys (left) and ImaGene Circles (right)

What is needed is a more specific background determination process that can account for the inherent variation between the gene and background regions. A fruitful avenue for such a reconstruction process could be that represented by the Texture Synthesis community. Efros et al. [68] proposed a non-parametric reconstruction technique that is now well established. The underlying principle of the work was to grow an initial seed pixel (located within a region requiring rebuilding) via *Markov Random Fields* (MRF). Although this works well, the nature of the approach is such that speed is sacrificed for accuracy. Bertalmio et al. [25] on the other hand took an approach that relies on the techniques as used by professional restorers of paintings; i.e, using the principle of an isotropic diffusion model. Chan et al. [39] extended these works along with other related techniques and proposed an elastic curvature model approach that combined amongst others Bertalmio's transportation mechanism along with the author's earlier *Curvature Driven Diffusion* (CDD) work to produce accurate yet relatively slow reconstructions.

Critically, the designs of all the reconstruction techniques are such that they work with natural imagery and therefore provide aesthetic reconstructions. Although the Oliveira et al. [155] technique was not designed for microarray images specifically, the authors tried to produce similar results to [25] albeit at a much faster pace. Alas, microarray images contain tens of thousands of regions requiring such reconstructions and are therefore computationally expensive to examine with the aforementioned techniques.

To address these problems, O'Neill et al. [157] attempted to harness ideas from the highlighted techniques and improve background prediction results while also reducing computation time somewhat. Specifically, O'Neill et al utilizes a simplification of the Efros et al technique in which gene spots are removed from the surface and recreated by searching known background regions and selecting pixels most similar to the reconstruction border. By making the new region most similar to given border intensities it is theorized that local background structures transition through the new region. However, the best such a process has accomplished in this regard is to maintain a semblance of valid intensities, while the original topological information is lost. The next section describes our novel approach that attempts to address some of these issues, e.g. retention of topology, process efficiency, and edge definition in a more natural way.

8.3 A new technique

In this work, we have proposed *Chained Fourier Image Reconstruction* (*CFIR*), a novel technique that removes gene spot regions from a microarray image surface. Although this may seem counter-intuitive (the gene spots are the elements of value in a microarray after all), the successful removal of these regions leads to more natural looking background surfaces, which can be used to yield yet more accurate gene spot intensities. Techniques as proposed by O'Neill et al. [157] work in the spatial domain exclusively and essentially compare all gene border pixels to those of the local background to produce probable pixel mappings. Although this works well, such brute force methods are typically computationally expensive. If we harness the frequency domain along with more traditional spatial ideas however we can render a reconstruction that deals with the issues (illumination, shading consistencies, etc.) more efficiently.

8.3.1 Description

The technique is designed to blend masked regions of the input image into their surrounding areas. So for example, a scratch on a much-loved photograph could be removed such that it is subdued significantly. In the context of this work, such a scratched artifact can be taken as the gene spot region itself. Therefore, removal of this "scratch" should yield the underlying background region in the gene spot area. However, due to the nature of the micro arraying process, gene spots can be rendered with different shapes and dimensions, individually and through the channel surfaces.

Therefore, a generic window centered at the target gene (as determined by GenePix) is used to capture all pixels $p_{x,y}$ within a specified square distance from this center. Note that (x, y) are the relative coordinates of the pixels in the window centered at pixel p. The Window size is calculated directly from an analysis of the underlying image along with resolution meta-data if needed. This window can then be used to determine the appropriate *srcList* and *trgList* pixel lists (foreground and background) accordingly. Note that in the current implementation, the window region resembles a square as defined by the outer edges of the diamonds in the left image of Figure 8.2. In a future implementation, this window shape could be determined dynamically such that larger artifact rich regions are accounted for more appropriately.

The gene spot pixels list can be defined via this windowed region as $G^p = \Omega^w(g_{x,y})$, with Ω^w representing pixels falling into the windowed region only and $g_{x,y}$ meaning such pixels should belong to the gene spot. The second list $B^p = \Omega^w(\bar{g}_{x,y})$ denotes those pixels within the same window that are not held in gene list G^p (and therefore must be representative of local background pixels).

With the two lists (*srcList* and *trgList*) in place, a Fast Fourier Transform (FFT) is applied to both lists independently. If $f(x, y)$ for $x; y = 0, 1, ..., M - 1; N - 1$ respectively denotes the $M \times N$ image region, the digital FFT for $F(u, v)$ can be defined as

$$F(u,v) = \sum_{M-1}^{x=0} \sum_{N-1}^{y=0} f(x,y)e^{-j2\pi\left(\frac{ux}{M} + \frac{vy}{N}\right)} \tag{8.1}$$

where (u, v) represent the frequency coordinates of their spatial (x, y) equivalents. Note the inverse transform is computed in much the same way. The real R, imaginary I and phase ϕ components of the resulting FFT spectrum can be broken up according to

$$|F(u,v)| = \left[R^2(u,v) + I^2(u,v) \right]^{\frac{1}{2}}, \text{ and} \qquad (8.2)$$

$$\phi(u,v) = tan^{-1} \left[\frac{I(u,v)}{R(u,v)} \right], \text{ respectively.} \qquad (8.3)$$

Global features of the image regions (repeating patterns, overall region intensity, etc.,) thus become localized within the frequency spectrum, while non-repeating structures become scattered.

Note it is crucial to retain the phase information as this has the effect of aligning the global features (as per the isotropic diffusion model of [25] for example) and therefore contains critical yet subtle surface characteristics (illumination and shading for instance). In order to capture this subtle intensity information within the background (*trgList*) region and allow the gene spot (*srcList*) area to inherit it, a simple minimization function can be used. Improved results can be rendered with the use of more complex matching criteria, but the minimum of the region was found to produce good results and is thus used at present.

$$R(u,v) = minimum \left| srcList(R), trgList(R) \right| \qquad (8.4)$$

The final stage of the algorithm replaces modified background (*trgList*) pixels within the gene spot (*srcList*) area with their original values. Recall, the FFT function disregards spatial information, which means subsequent modifications (like the minimizer function) could well change inappropriate pixels with respect to the reversed FFT. Therefore, the original non-gene spot pixels must be copied back into the modified regions such that erroneous allocations are not propagated through the reconstructed region during the next reconstruction cycle.

8.3.2 Example and pseudo-code

The CFIR process as outlined previously can best be explained with the use of a walk-through example. Initially, the process creates two distinct lists for a given gene spot location. One list represents the gene spot pixels themselves as demarcated within a square window centered at the gene, while the other consists of the pixels assigned to this gene spot's local background within the window. The discrete equations of Section 8.3.1 are then brought to bear on these lists with the local background taken as the source image and the gene pixels the target or reconstructed image. In essence, we are capturing the intensity information from the background region and modifying the foreground pixels accordingly. Figure 8.3 presents a sample-reconstructed region

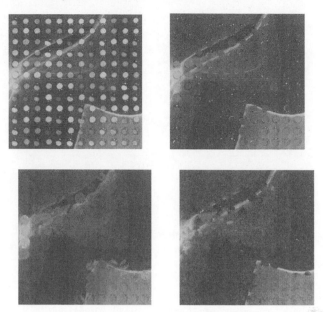

FIGURE 8.3
Reconstruction examples: original image (top left), reconstructed GenePix
(top right) O'Neill (bottom left) and CFIR (bottom right) regions

from the Figure 8.1 image as processed by the techniques. Although the application of spatial methods when applied separately to such reconstruction problems can work well, there are better ways to carry out such tasks. The formulation as described for CFIR allows us to combine advantages from both the frequency and spatial domains such that the reconstructed regions retain implicit domain information and reduce time complexity over contemporary methods.

A pseudo-code implementation for the CFIR algorithm can be found in Algorithm 8.1. For clarity, this implementation has been based around the processing of individual target windows with two explicit groups containing either signal or noise elements.

```
1  Inputs
2     srcList: List of gene spot region pixels
3     trgList: List of sample region pixels
4  Outputs
5     outList: srcList pixels recalibrated into trgList range
6  Function fftEstimation(srcList,trgList):outList
7  For each gene
8     srcMask=fourier transform srcList members
9     trgSample=fourier transform trgList members
10    recon=srcMask*trgSample // generates initial reconSurface
11   While doneIterate=0
```

```
12   recon=fourier transform initial reconSurface
13   reconPhase=phase elements of reconSurface
14   minimum recon element=smallest element in trgSample surface
15   recon=inverse fourier merged with recon and reconPhase
       surfaces
16   // retains subtle surface characteristics
17   recon elements _0=smallest element in srcMask
18   recon elements _65535=largest element in srcMask
19   reset non-gene pixels in recon to trgList
20   if difference between recon and trgSample _tolerance
21   doneIterate=1
22   End If
23   End While
24   outList=recon region
25 End For
26 End Function
```

Algorithm 8.1: Chained Fourier Reconstruction Functions Pseudo-Code

8.4 Experiments and results

This section details numerous experiments designed to test empirically the performance characteristics between the O'Neill and CFIR algorithms with respect to GenePix. Although there are many ways that such performance characteristics can be distilled, in this work the focus is on the median expression intensities. These intensities become the raw gene expressions (as used in post-analysis [44, 70, 92, 120, 167, 168] work for example) and therefore garner the ability to provide insight into a reconstruction event. In particular, such values help provide greater understanding into a gene spot's repeat set and therefore assist with clarification of the quality of the reconstruction event.

8.4.1 Dataset characteristics

The images used in this chapter derive from the human gen1 clone set data. These experiments were designed to contrast the effects of two cancer-inhibiting drugs (PolyIC and LPS) over two different cell lines. One cell line represents the control (or untreated) and the other the treatment (HeLa) line over a series of several time points. In total, there are 47 distinct slides with the corresponding GenePix results present. Each slide consists of 24 gene blocks with each block containing 32 columns and 12 rows of gene spots. The gene spots in the first row of each odd-numbered block are known as the Lucidea Score-Card [178] and consist of a set of 32 pre-defined genes that can be used to test various experiment characteristics, binding accuracy for example. The remaining 11 rows of the odd-numbered blocks contain the human genes themselves. The even-numbered blocks are repeats of their odd-numbered counterparts. This means that each slide has 24 repeats of the 32 ScoreCard genes and 4224 repeats of the human genes respectively. Note it is generally accepted that extreme pixel values should be ignored as these values could go beyond the scanning hardware's capabilities.

8.4.2 Synthetic data

The technique relies on the assumption that a foreground region can be removed from the image such that the identification of it will be lost after the construction event. In order to verify this "lost" ability, one would need to rebuild an obscured known region and compare the before and after surfaces for accuracy. However, it is not possible to determine the optimal background pixels under a given gene spot region as the gene spot is in the way. Therefore, we take a slightly different approach to this verification stage by creating *Synthetic Gene Spots* (SGSs), which will be placed into known background areas. Then, the reconstruction process can be executed on these synthetic regions as normal. Because we already have the before shot of the background area, the reconstructed SGS region can be marked correctly for effectiveness. Such an absolute comparison renders a clear understanding of the algorithm's reconstruction characteristics as well as the potential flux error therein.

The first experiment was therefore designed to determine how well a reconstruction process could remove a synthetic gene from the image. The problem now would be how to generate appropriate synthetic regions to begin with. In order for an SGS to be representative of a true gene spot, it must contain some semblance of the original characteristics. In essence, such characteristics can be broken up into the following three aspects:

8.4.2.1 Semi-transparent

Gene spots tend to have a semi-translucent surface due mainly to the way they are created/deposited onto the glass slide. The translucency results in the gene spot's background element blending into the gene. Such blending is accounted for (in part) by the quantification processes background subtraction stage.

8.4.2.2 Uniform surface

Although a typical microarray images gene spots are quite different in intensity from their outer-lying background regions, their internal composition looks to be very close to constant. However, with the appropriate removal of the transparency component, we would find that a gene spot's composition tends to fluctuate more so than first envisioned.

8.4.2.3 Chemical propensity

The chemical composition of the two or more color dyes (Cy5 and Cy3 in this context) used to track the binding of the underlying experiments are known to have strength biases. That is to say, Cy5 tends to permeate the surface more actively than that of Cy3. This leads to the situation where the Cy5 surface can hide weaker Cy3 regions. This phenomenon is more critical in weak gene spots, where the Cy3 region gene is essentially lost to background clutter.

With use of the characteristic information, it is now possible to ascertain the biological effect over microarray imagery. The test experiment used 64 realistic SGSs placed into existing background regions of the Figure 8.1 images Cy5 and Cy3 surfaces. This effect yields a ball-park-figure for the potential distillation errors generated by the various background reconstruction techniques. Such distillation potentials as rendered from the test imagery can be seen in Figure 8.4.

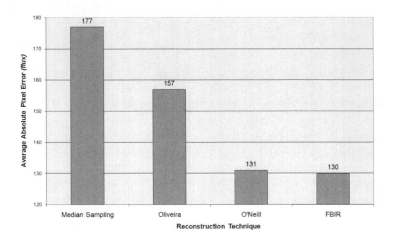

FIGURE 8.4
Synthetic gene spots: average absolute pixel error

The graph shows that on average, the GenePix advocated median sampling approach to background reconstructions yield a potential intensity error of 177 flux per pixel per SGS region. The other techniques yield smaller error estimates. A consequence of such a finding reiterates that downstream analysis based on GenePix (specifically the BackGround Correction (BGC) stage) estimates directly produce more erroneous gene expressions than perhaps realized.

8.4.3   Real data

With potential error types highlighted for the GenePix background generation technique, the next stage of testing involved determining how these errors translate into the reconstruction of real gene spot regions.

Experiment two therefore was interested in examining the impact that the reconstruction techniques have on the artificial or ScoreCard control genes for all blocks across the test set. Recall, the composition of the test imagery is such that we have "24 repeats of 32 control genes" rather than "2 repeats of 32 human genes." In addition, as the control genes are completely independent of the biological experiment, they should fluoresce in exactly the same way (ideally) across all images regardless.

Standard deviations for the 32 ScoreCard control genes are thus analyzed across the 24 repeat locations of the 47 experiments. The underlying assumption being that the variances of these genes should be lower when better reconstruction techniques are used. In turn, this would mean the techniques account for BGC more appropriately. The Figure 8.5 plot show this tracking process with respect to ScoreCard genes.

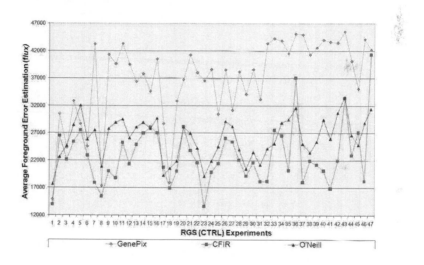

FIGURE 8.5

Real gene spots: overview of ScoreCard gene standard deviations for the 47 experiments

Although the GenePix process plot looks to be partially smoother than the other reconstruction techniques, the magnitude of the plot is such that it contains greater error potential. The CFIR and O'Neill process plots on the other hand have reduced the potential error and therefore are more representative of valid expressions measures.

The third experiment conducts the performance evaluation of CFIR, O'Neill and GenePix. Here, the focus is on the relationships between expression measurements for all ScoreCard genes in the same slide (the top figure in Figure 8.6) and across all slides in the test set (the bottom figure in Figure 8.6). In all cases the underlying assumption for repeated genes is that they (the genes) should have highly similar intensity values (ideally) for a given time point, regardless of their location on the slide surface. We would though expect to see some differences in intensity measures as the time point's increase through the duration of the biological process.

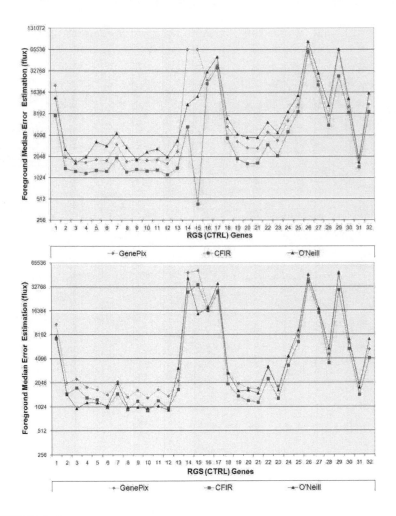

FIGURE 8.6

Real gene spots: absolute medians for 32 genes over Figure 8.1 (top figure) and test set (bottom figure) regions

The plots represent the absolute foreground median from both image channels for the techniques. From the top figure in Figure 8.6 it is clear that CFIR outperformed the other methods comfortably. Note, however, how saturated gene spots - number 15 for instance cause a blip in the profile plot. By their very nature, the surface of a saturated gene is close to a constant value and obviously artificially high. During the reconstruction stage, such high values differentiate from their neighboring pixels significantly. This has the effect of generating a bigger reduction in the resulting gene surfaces measure accordingly. The final residuals for the reconstructed techniques are 14514, 7749, and 13468 flux for GenePix, CFIR and O'Neill respectively. In this case, the saturated gene spots did not affect the outcome of the final quantification stage a great deal. Indeed, as far as control data is concerned the CFIR process reduced noise by ~46%.

The top figure in Figure 8.6 plot does however represent a specific image surface, which cannot impart how reconstruction techniques fair across a range of image qualities in a foreground sense. Recall, the plots of Figure 8.5 only showed the reduction possible in a surface's profile sense. The question here then is how exactly such a reduction manifests itself onto the final gene metrics. Therefore plotting the entire 47-slide test set cross-sectional average data highlights such information.

The O'Neill and CFIR processes are much closer overall as compared to the sample slide; however, there are subtle handling differences in the saturated gene regions. Saturated genes in the general case fair better with CFIR than they do with O'Neill. The respective residuals for the test set are 10374, 7677 and 9213 flux respectively. This is indicative that CFIR tends to produce lower scores generally.

It is clear that reconstructing the gene spot background does have a positive effect on the final expression results (as indicated by the previous figures) but, not so obvious, are the ramifications this reconstruction has over the test set. The top figure in Figure 8.7 therefore plots a comparison chart which shows explicitly the improvement (or not) of a particular reconstruction method as compared to the original GenePix expressions. Additionally, execution time plays a critical role in the reconstruction task, as techniques need to run as fast as possible given the number of gene spots that must be processed. Therefore, the bottom figure in Figure 8.7 presents a brief breakdown of the timings required for the techniques to parse a small percentage of the test set.

The distinct banding occurring in gene regions 3~8 and 16~19 of the top figure in Figure 8.7a are associated with saturated (or near background) intensities as created by the scanner hardware and suggest more work is needed with respect to these genes. The non-banded genes on the other hand are indicative of the individual reconstruction techniques being able to account more appropriately for gene intensity replacement. The table of Figure 8.7 highlights the significant speed increase gained by applying the CFIR process to reconstruction over the O'Neill and GenePix methods.

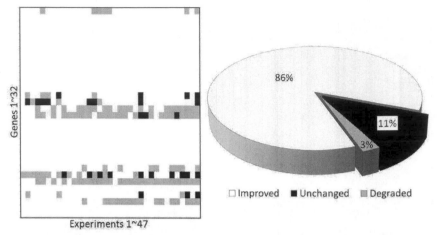

	Time on Xeon 3.4GHz (hh:mm:ss)		
Image	_GenePix_	_O'Neill_	_CFIR_
Hela PolyIC t00_(1a)_	4:00:00+	1:53:00	00:52:39
Hela PolyIC t05	4:00:00+	1:58:00	00:58:15
Hela PolyIC t18	4:00:00+	1:52:00	00:54:44

FIGURE 8.7

Final results comparison: matrix for test set showing difference in repeat expression fluctuations (top figure); GenePix, CFIR and Both techniques are assigned the colors black, white and grey (~10% difference) respectively and Sample Timing Chart (bottom figure)

8.5 Conclusions

The chapter looked at the effects of applying both existing and new texture synthesis inspired reconstruction techniques to real-world microarray image data. It has been shown that although the use of existing methods can be highly effective, their output quality and execution time with respect to microarray data needs to be significantly improved.

To overcome such accuracy and timing challenges, we proposed a novel approach to reconstructing a gene's underlying background by attempting to harness an image's global knowledge more intently along with the gene's neighbor pixels. The proposed technique takes advantage of the grouping concept of the frequency domain and characterizes the global entities in a systematic way. At the same time, a gene spot's local spatial knowledge is used to help restrict the spread of intensity within the constructed region. The results as obtained from the study have shown significant improvements over a commonly used package (GenePix) and a brute force approach (O'Neill). Specifically, not only was the gene repeat variance reduced from slides in the test set, but construction time was also decreased about 50% in comparison to O'Neill, and more than 75% to GenePix.

In the future, we would like to compare against other mainstream analysis methods like Spot [227], ScanAlyze [183], ImaGene and QuantArray [169] for instance. In addition, as hinted at with the saturated genes, sub-allocation of replacement pixels can be problematic in cases where said genes straddle strong artifact regions. Whereas a weighted transition map could prove beneficial to the sub-allocation problem, we believe that the current CFIR reconstruction process could also be enhanced by using a directional Gaussian grating process during iterations. Another potential path could be the use of a wavelet transforms based approach rather than raw FFTs.

9

Graph-Cutting for Improving Microarray Gene Expression Reconstructions

9.1 Introduction

Although microarray technology [186] was invented in the mid-1990s the technology is still widely used in laboratories around the world today. The microarray "gene chip" contains probes for an organism's entire transcriptome where differing cell lines render gene lists with appropriate activation levels. Gene lists can be analyzed with application of various computational techniques, be they clustering [70] or modeling [119] for example such that the differential expressions can be translated into a clearer understanding of the underlying biological phenomena present. For a detailed explanation of the micro arraying process readers may find references [47, 67, 133, 166] of interest.

Addressing the issue of microarray data quality effectively is a major challenge, particularly when dealing with real-world data, as "cracks" will appear regardless of the design specifications etc. These cracks can take many forms, ranging from common artifacts such as hair, dust, and scratches on the slide, to technical errors like miscalculation of gene expression due to alignment issues or random variation in scanning laser intensity. Alongside these errors, there exists a host of biological related artifacts such as contamination of the complementary Deoxyribonucleic Acid (cDNA) solution or inconsistent hybridization of multiple samples. The focus in the microarray field therefore is on analyzing the gene expression ratios themselves [44, 70, 92, 120, 167, 168] as rendered from the image sets. This means there is relatively little work directed at improving the original images [88, 89, 157, 227] such that final expressions are more realistic.

As noise in the images has a negative effect with respect to the correct identification and quantification of underlying genes, in this chapter we present an algorithm that attempts to remove the biological experiment (or gene spots) from the image. In the microarray field, it is accepted as part of the analysis methodology that the background domain (non-gene spot pixels) infringes on the gene's valid measure and steps must be taken to remove these inconsistencies. In effect, this removal process is equivalent to background reconstruction and should therefore produce an image which resembles the "ideal" background more closely in experimental (gene spot) regions. Subtracting this new background image from the original should in-turn yield more accurate gene spot expression values. The gene expression results of the proposed reconstruction process are contrasted to those as produced by GenePix [14] (a commercial system commonly used by biologists to analyze images). Results are also compared with three (3) of the aforementioned reconstruction approaches (O'Neill et al. [157], Fraser et al. [88,89]) with respect to like-for-like techniques.

The chapter is organised in the following manner. First, we formalize the problem area as it pertains to microarray image data and briefly explain the workings of contemporary approaches in Section 9.2. Section 9.3 discusses the fundamental idea of our approach with the appropriate steps involved in the analysis highlighted. We then briefly describe the data used throughout the work and evaluate the tests carried out over both synthetic and real-world data in Section 9.4. Section 9.5 summarizes our findings and renders some observations into possible future directions.

9.2 Existing techniques

Microarray image analysis techniques require knowledge of a given gene's approximate central pixel and the slide's structural layout; therefore, all analysis techniques have similarities (regardless of their specific implementations). For example, a boundary is defined around the gene - thus marking the foreground region - with any pixels outside a given radius taken to be local background. The median value for this background is then subtracted from the foreground and the result is summarized as a $\log_2$ ratio. Bounding mechanisms include partitioning pixels via their histograms [44, 169], edge-based [9, 165], region growing [1,9] and clustering [118,139] functions, a detailed comparison of the more common approaches can be found in Yang et al. [227].

The underlying assumption throughout these mechanisms however is that there is little variation within the gene and background regions. Unfortunately this is not always the case as can be seen in the example regions of the top figure in Figure 9.1, which depicts a typical test set slide (enhanced to show gene spot locations) with a total of 9216 gene regions on the surface held within an approximate area of $\sim 5000 \times 2000$ pixels. Note in addition that every image in the test set was created on a so called two-dye microarray system which means the DNA tagging agents used for the two-channels are known as Cyanine 5 (Cy5) and Cyanine 3 (Cy3). The close-up sections provide good examples of the low-level signal produced in the image; problems such as partial or missing gene spots, shape inconsistencies, and background variation are clearly evident. In particular, note how the scratch and background illuminations around the genes change significantly. Note that all figures and diagrams are best viewed in color.

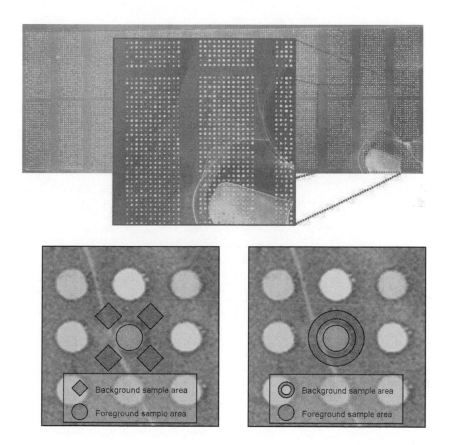

FIGURE 9.1

Example images: typical test set slide illustrating structure and noise (top figure) with sample gene, background locations for GenePix Valleys (bottom left) and ImaGene Circles (bottom right)

A background identification process is required such that inherent variations between gene and background regions are handled more appropriately. Texture Synthesis represents one possible avenue for such reconstruction approaches as they deal with a similar problem. For example, Efros et al. [68] proposed a non-parametric reconstruction technique that is now well established. The underlying principle of the work was to grow an initial seed pixel (located within a region requiring rebuilding) via Markov Random Fields (MRF).

Bertalmio et al. [25] took an approach inspired by the techniques as used by professional restorers of paintings; i.e. the principle of isotropic diffusion, to achieve their reconstruction. Inspired by this work, Oliveira et al. [155] attempted to produce similar results, albeit much faster. Indeed, Oliveira traded accuracy for speed and succeeded in reducing computation complexity somewhat with their results being of a similar level. Chan et al. [39] then extended these works along with other related techniques and proposed an elastic curvature model approach that combined amongst others Bertalmio's transportation mechanism with the authors earlier Curvature Driven Diffusion models to produce yet more accurate reconstructions. Sun et al. [200] proposed an interactive inpainting method with missing strong visual structures by propagating structures (similar to the Bertalmio approach) according to user-specified curves, such an approach does improve on previous methods somewhat, but the interactivity clause would be inappropriate in the context of microarray data (due to the number of objects that need to be rebuilt therein).

In 2003 O'Neill et al. [157] attempted to address the issue of applying inpainting methodologies into object removal (for microarray imagery specifically) by harnessing ideas from the Efros et al. technique. Specifically, O'Neill et al. remove gene spots from the surface by searching known background regions and selecting pixels most similar to the reconstruction border. By making the new region most similar to given border intensities it is theorized that local background structures transition through the new region. However, the best such a process has accomplished in this regard is to maintain a semblance of valid intensities, while the original topological information is lost. This is not to say however that such an approach is void of merit as the resulting surface reconstructions do significantly improve on prior methods ([227] and Axon [14] for example). In addition, although O'Neill et al. moved away from the inherent aesthetic bounds of the parent techniques to generate more accurate background pixel estimates, it is perhaps unsurprising that they found a blur process to be beneficial for their final results.

An alternative approach to the inpainting mechanism that could be of great benefit in this medical image context is that as held in the Graphing community. Specifically, Graph-cutting type processes as used in the general field of image-editing could have potential with respect to such a reconstruction problem. For example, Perez et al. [164] proposed a Poisson Image Editing technique to compute optimal boundaries between source and target images, while Agarwala et al. [4] created an interactive Digital Photomontage system that combined parts of a set of photographs into a composite picture. Kwatra et al. [124] proposed a system that attempted to smooth the edge between different target and source images and is perhaps closest to our current ideology. Other Graph related methods can be seen with References [20, 153, 174] for example. The critical point with respect to these approaches is that they contain an interactive element. Clearly the interactive element lends itself to an aesthetic resultant very well as such aesthetic improvements are subjective in nature.

However, microarray images (and indeed medical imagery generally) contain tens of thousands of regions requiring such reconstructions and are therefore either computationally expensive to examine with the aforementioned techniques (not practical) or their interactivity clause renders them to be of limited use (with respect to the number of objects to process). Also, let us not forget that such methods are focused at aesthetic reconstructions. Medical images by their very nature demand reconstruction processes that go beyond such aesthetic considerations.

What is needed is a technique that generates true pixel replacements for an area needing rebuilding rather than the estimates as returned by current approaches. It would also be beneficial if the technique only rebuilt regions that required it (meaning the bounding box around a gene did not include unnecessary pixels as per O'Neill). The next section describes an approach that attempts to address some of the issues related to object removal by using a graph detection mechanism in an automatic and natural way.

9.3 Proposed technique

In this chapter, we propose Graphs-Cut Image Reconstruction (*SCIR*), a technique that removes gene spots from a microarray image surface such that they are indistinguishable from the surrounding regions. Removal of these regions leads to more accurate gene spot intensities. Our previous work in this domain examined the effects of Recalibration (*HIR*) and Fourier Chaining (*CFIR*) (Fraser et al. [88, 89] respectively) techniques. Although *CFIR* dealt with shading and illumination issues more appropriately than *HIR*, *HIR* produced similar results significantly faster. However, both techniques can produce poorer reconstructions in regions dominated by strong artifacts (a saturated gene surrounded with a similar level artifact for example). This work therefore attempts to improve on this issue; while at the same time generating exact pixel values.

9.3.1 Description

The technique is designed to replace gene spot pixels with their most appropriate background neighbor. For example, a scratch on a photograph could be removed such that it is unidentifiable after reconstruction. In the context of this work, a scratch is equivalent to the gene spot region itself. Therefore, removal of this "scratch" should yield the underlying background region in the gene spot area. However, due to the nature of the micro arraying process, gene spots can be rendered with different shapes and dimensions, individually and through the channel surfaces.

Therefore, we use a window centered at a target gene (as determined by GenePix) to capture all pixels $p_{x,y}$ within a specified square distance from this center. Note x, y are the relative coordinates of the pixels in the window centered at pixel p. The Window size is calculated directly from an analysis of the underlying image along with resolution meta-data. The window can then be used to determine the appropriate *srcList* (list of gene spot region pixels) and *trgList* (list of sample region pixels) pixel lists respectively.

The gene spot pixels list can be defined via this windowed region as, $G^p = \Omega^w(g_{x,y})$, with Ω^w representing pixels falling into the windowed region and $g_{x,y}$ meaning those pixels falling into the gene spot. The second list $B^p = \Omega^w \bar{g}_{x,y}$ denotes those pixels within the same window that are not held in gene list G^p (and must therefore be representative of local background pixels).

The Graph-Cutting process then uses the *srcList* to determine those neighboring pixels that have the strongest intensity through the surface. While *trgList* is used to determine the weakest neighboring background intensities respectively. In the general sense, if we let image $\mathbf{I}$ be a $n \times m$ surface, and if $x, y = 0, 1, ..., M - 1; N - 1$ parses said image, a vertical graph cut Gv through the two lists could be defined as

$$Gv = \{Gv_v\}_{x=1}^n = \{(x, Gv(x))\}_{x=1}^n, \forall x, |Gv(x)\text{-}Gv(x\text{-}1)| \leq m, \qquad (9.1)$$

The vertical graph therefore is an 8-way connected neighboring set of pixels in the image from top-to-bottom with one pixel per row. Initially, the image is parsed such that cumulative energy for all possible connected pixel sets should be at a minimum for each x, y pairing through the surface.

In essence a mapping of this nature means that strong foreground pixels are replaced with their appropriate weak background equivalents. Such a replacement policy guarantees that the new foreground surface is not artificially biased to a particular intensity range. Indeed, if anything the new regions will consist of slightly lower intensity than perhaps is necessary meaning therefore a built-in buffer is also applied presently.

9.3.2 Pseudo-code and example

A pseudo-code implementation of the SCIR algorithm can be found in Algorithm 9.1. For clarity, the implementation is based on processing target window regions, which each contain a distinct set of pixels that are separated into gene spot and background sets.

```
1 Inputs
2     srcList: List of gene spot region pixels
3     trgList: List of sample region pixels
4 Outputs
5     outList: srcList pixels recalibrated into trgList range
6 Function graphsCut(srcList,trgList):outList
7 For each gene
8     geneRadius=radius of current gene spot
9   While geneRadius <= 0
```

```
10    fgEnergy=max pixel surface from srcList members
11    bgEnergy=min pixel surface from trgList members
12    fgChain=Parse fgEnergy to determine max-neighbor pixel
         chain
13    bgChain=Parse bgEnergy to determine min-neighbor pixel
         chain
14    remove fgChain from fgEnergy
15    remove bgChain from bgEnergy
16    copy bgChain pixels into srcList locations
17    geneRadius-=1
18    outList=srcList
19   End While
20 End For
21 End Function
```

Algorithm 9.1: Graphs-Cut Reconstruction Functions Pseudo-Code

Initially, the SCIR process creates two distinct lists for a given gene spot location. The source list represents gene spot pixels as demarcated within the square window centered at the gene, while the target list consists of the remaining pixels in the window. Equation (9.1) is executed on the lists with the local background taken as the source region and the gene pixels the region to be reconstructed. Essentially the approach tries to create a chain (or neighboring set) of pixels through the region that have (in some sense) a maximal/minimal intensity respectively. This can be thought of as a gradient function that searches for high-contrast (or edge) pixels within the gene spot region and low-contrast pixels within the local background region.

Figure 9.2 presents a sample-reconstructed region from the Figure 9.1 image as processed by the documented techniques.

Note in particular how the SCIR surface looks sharper than that of O'Neill. This is due in part to the O'Neill surface being blurred such that resulting outliers, etc., are suppressed. The SCIR technique on the other hand generates absolute surfaces without this blur stage.

9.4 Experiments and results

This section details numerous experiments that were designed to empirically test the performance characteristics of the reconstruction methods. Median expression intensities are utilized in the comparisons as these values are infact the raw gene expressions (as used in post-analysis [44,70,92,120,167,168] work for example). These values help provide clearer understanding of a gene spot's repeat set and as such assist with clarification of the reconstruction quality itself.

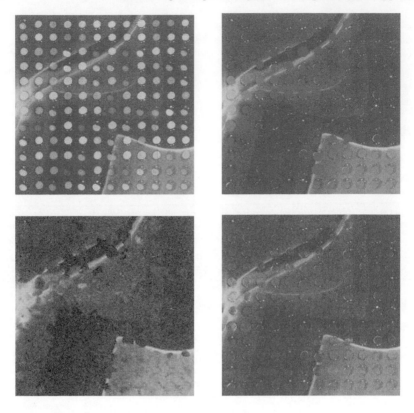

FIGURE 9.2
Reconstruction examples: original image (top left), reconstructed GenePix (top right) O'Neill (bottom left) and SCIR (bottom right) regions

9.4.1 Dataset characteristics

The dataset characteristics is the same as that given in Section 8.4.1.

9.4.2 Synthetic data

The guiding principle of the technique is the feasibility that replacing gene spot pixels with pixels from neighboring regions will result in a reconstructed area that is indistinguishable from the neighboring region. Put another way, the gene spots should simply vanish from the surface which means that their new texture has to be very similar to the neighboring region. Note that regions with strong and sharp intensity differences (an artifact edge for example) will be harder to "blend" successfully. In order to verify that the principle is at least valid, one would need to rebuild an obscured known region and compare before and after surfaces for accuracy. However, as the gene spot sits above the optimal background surface it is not possible to determine optimal rebuild pixels. In order to validate rebuild feasibility therefore, we use the *Synthetic Gene Spot* (SGS) creation process as outlined in Fraser *et al.* [89].

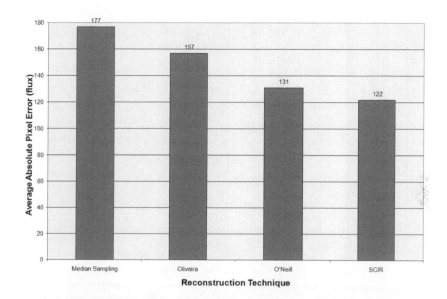

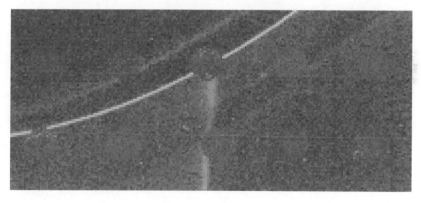

FIGURE 9.3
Synthetic gene spots: average absolute pixel error (top figure) and close up of Figure 8.1 region with ten synthetic spots (bottom figure)

The first experiment is focused at answering "how well the SCIR process removes synthetic gene spots from the image"? Sixty-four (64) realistic SGSs were placed into existing background regions of the Figure 9.1 images Cy5 and Cy3 surfaces. These synthetic genes were then reconstructed with the before

214 Microarray Image Analysis: An Algorithmic Approach

and after surfaces compared for similarity. Note that as the artifact region itself could be considered gene spot similar, our reconstruction processes also attempt to build the region such that the artifact pixels are removed. This process yields a ball-park-figure for the potential distillation errors generated by the various background reconstruction techniques. Such potentials as rendered from test imagery can be seen in the top figure in Figure 9.3, while the bottom figure of Figure 9.3 highlights a close up sample region of the aforementioned SGSs.

The graph presents the potential intensity flux error (PIFE) for the reconstruction techniques. On average, the GenePix advocated median sampling approach yields a PIFE of 177 per pixel per SGS region while the other techniques yield decreasing values (our process value of 122 represents a ~30% reduction over GenePix). Such a finding reiterates that downstream analysis when based on GenePix (specifically the BackGround Correction [BGC] stage) estimates directly, produce more erroneous gene expressions than perhaps appreciated.

The panel (bottom figure of Figure 9.3) surface highlights a sample of the SGS region with a large artifact running through two gene spot regions. Also note we can see that the strong artifact edge has been successfully replaced with appropriate background substitutions. Note, however, that such strong edges can cause greater challenges within real data as shall be seen.

9.4.3 Real data

With our confidence in the reconstruction techniques abilities enhanced by the synthetic results, the next stage is to understand how such reconstructions fare with real data. In particular, "how badly do strong artifact edges interfere with a reconstruction event"?

Experiment two only uses the ScoreCard control genes for all blocks across the test images. Recall, the composition of the test imagery is such that we have more technical repeats of the control genes than the human ones. Also, the control genes are completely independent of the biological experiment which means ideally they should fluoresce in exactly the same way across the images regardless of environmental conditions (in principle).

The Figure 9.4 plot presents the tracking of the standard deviations (STD) for the 32 ScoreCard genes over the 24 repeat locations. Note, however, that due to the way in which O'Neill calculates a given gene spots region, their STD's are somewhat lower than expected. However, the plot still imparts general characteristics for the given reconstruction techniques.

If we disregard saturated gene spots for a moment and examine the close up section of the plot we see the profile residuals follow each other fairly well. This means the processes do reduce STDs at least in a partial sense.

Critically, then, this leads to a need to understand the reconstruction technique's performance characteristics more closely. Specifically the relationships between expression measurements for all ScoreCard genes in the same slide

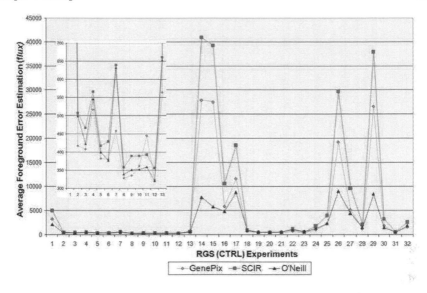

FIGURE 9.4

Real gene spots: overview of ScoreCard gene standard deviations

(top figure of Figure 9.5) and across all slides in the test set (bottom figure of Figure 9.5) are compared. Note that it is expected that some intensity differences will appear as the experimental time points increase as required through the biological processes.

These plots show the bound absolute foreground median values for the multiple image channels for the documented techniques. From the top figure of Figure 9.5 it can be seen that SCIR and O'Neill performed in a similar vein with very little difference amongst them. However, the saturated gene spots - 15 in this case has caused a blip in the profile plot for SCIR. Recall, that by the very nature of a saturated gene spot, the surface is close to a constant value and obviously artificially high. But in this instance the gene in question also has a strong artifact intercepting it. During reconstruction, the constant type value of the gene is not a major challenge to rectify; more problematic is how to deal with the strong intercepting artifacts appropriately. Note that the replacement pixel sets as derived during reconstruction actually do a fair job overall. For this image, the saturated gene spots did not affect the outcome of the final quantification stage greatly.

Whereas the top figure of Figure 9.5 plot represents a specific image surface, which does not render a given reconstruction technique's abilities to deal with a range of image modalities. The bottom figure of Figure 10.5, therefore, shows the same information across the entire 47-slide test set. This should allow us to see how exactly a reduction manifests itself onto the final gene metrics. Clearly, the SCIR process has reduced the technical repeats to a

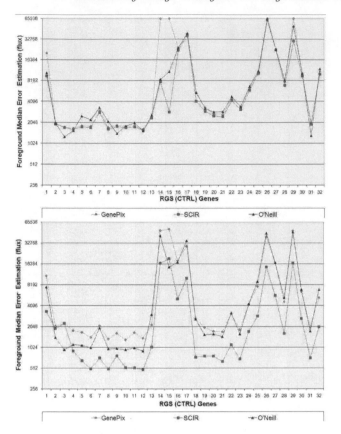

FIGURE 9.5

Real gene spots: absolute medians for 32 genes over Figure 8.1 (top figure) and test set (bottom figure) regions

greater extent than perceivable from the sample image alone. The respective profile values for the test set are 10374, 3742, and 9213 flux respectively.

Clearly, reconstruction of gene spot's does have a positive effect on the final expression results but, not so obvious, are the ramifications that the reconstruction has over the test set. The left panel of Figure 9.6, therefore, is a comparison chart showing explicitly the improvement (or not) of a particular reconstruction technique against the original GenePix expressions.

The general banding region of genes 16~17 and partial banding of gene 30 as seen in Figure 9.6 are associated with aforementioned saturated (or near background) gene intensities as created by the Axon [14,15] scanner hardware and are suggestive of more work needed. The non-banded genes on the other hand are indicative of the individual reconstruction techniques being able to account more appropriately for gene intensity replacement.

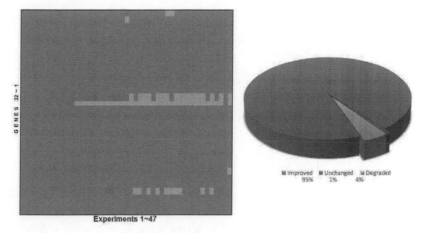

FIGURE 9.6

Final results comparison: matrix for test set showing difference in repeat expression fluctuations; the GenePix, SCIR and O'Neill techniques are assigned the colors blue (darkest), red and green (lightest) (~10% difference) respectively

9.5 Conclusions

The chapter looks at the effects of applying current and new texture synthesis inspired reconstruction techniques to real-world microarray image data. In particular, we propose an approach to reconstructing a gene's underlying background by attempting to focus on problematic pixels only. Previously, we took a purely local constraint based approach to the problem and, although the reconstructions were better than those of contemporary approaches, clear areas of improvement existed. Later work relaxes such local constraints and tries to use a level of "localness" to guide the harnessing of an image's global knowledge more closely. Although such a holistic process was shown to be highly effective and a great improvement, again the approach still had weaknesses. The primary weakness is that as related to the nature of the reconstruction task as, in order to rebuild a region appropriately a local area must be sampled in some way. Such a sampling must be larger than the actual region to be rebuilt as local gene spot extended edge discontinuities need to be taken into account. However, not only does such a consideration increase computation complexity but, one could argue that rebuilding the extended edges causes an artificial increase in flux error which is propagated through to final expressions. Note an extended edge type problem exists in a gene spot's internal regions also.

In this work, we were specifically interested in addressing issues related to extended edge problems of gene spot reconstruction. The proposed approach therefore utilizes a graph theory inspired pixel identification mechanism to select those pixels that are most similar to their direct neighbors within a pre-defined region. The highlighted pixel chains are then replaced according to their closest background region representative. The results show that the new method makes a significant improvement in gene expression reduction both directly and when compared with technical repeat variances.

Although in the future we would like to be able to broaden our analyses with respect to contrasting our reconstruction processes with other mainstream methods like Spot [227], ScanAlyze [183], ImaGene, and QuantArray [169], it is difficult to acquire appropriate final results as our collaborators use GenePix exclusively. In addition, a critical element of such a critique would require the internal result workings of the mentioned methods.

It is quite probable that a hybrid reconstruction system (able to classify to some extent a gene region) will be of great benefit to this analysis task. Such a hybrid system would use what is deemed to be the most appropriate reconstruction technique for a given gene. As we have now developed several separate reconstruction techniques, shown to be highly effective at their task, it is our belief that such a hybrid system can now be tackled appropriately as several reconstruction specific component parts are in place.

10

Stochastic Dynamic Modeling of Short Gene Expression Time Series Data

10.1 Introduction

DNA microarray technology has provided an efficient way of measuring the expression levels of thousands of genes in a single experiment on a single "chip." It enables the monitoring of expression levels of thousands of genes simultaneously. The potential of such technologies for functional genomics is tremendous. Measuring gene expression levels in different conditions may prove useful in medical diagnosis, treatment and drug design. Microarray technology has been heralded as the new biological revolution after the advent of the human genome project, since it become possible to extract the important information from gene expression time-series data.

Through the extensive microarray image stages conducted in previous chapters, we would have obtained a global view on the expression levels of all genes when the cell undergoes specific conditions or processes. Obviously, the next step is to infer useful biological information and determine the relationships between individual genes. In this regard, many current research efforts have focused on clustering. Cluster analysis of the gene expression data appeared first in [70] and has quickly attracted considerable research attention. A number of clustering algorithms have been examined on gene expression data, such as hierarchical clustering [70], self-organizing map [202], k-means [203] and Gaussian model-based clustering [171, 228], to name just a few [110]. However, a fundamental shortcoming of such clustering schemes is that they are based on the assumption that there exists the correlation similarity between genes. Recently, there has been an increasing research interest to reconstruct models for gene regulatory networks from time series data [56, 190], such as Boolean network model [5, 106, 131, 194], linear differential equation model [43, 55, 58, 105], Bayesian model [96, 119, 132, 146], state space model [22, 173, 222] and stochastic model [48, 207].

Obviously, selecting a good model to fit gene regulatory networks is essential to a meaningful analysis of the expression data. It turns out that the model for gene regulatory networks should posses the following three properties. First, the model should be easy to evolve the biological information, such as the linear dynamical model. Second, the model should reflect the "stochastic" characteristics, since it is well known that the gene expression is an inherently stochastic phenomenon [121, 136, 146, 204]. Third, the observations (measurement outputs) of the model should be regarded as noisy due to our inability to perfectly and accurately (noise-free) measure gene expression levels. Fourth, in biology and medicine, the available time series (e.g., gene expression time series) typically consists of a large number of variables but with a small number of observations. Therefore, the modeling method should be capable of tackling short time series with acceptable accuracy.

There have been attempts to reconstruct models for gene regulatory networks by taking into account the aforementioned three properties. Dynamic Bayesian networks have been proposed to model gene expression time series data [119, 132, 146]. The merits of dynamic Bayesian networks include the ability to model stochasticity and handle noisy/hidden variables. However, dynamic Bayesian networks need more complex algorithms such as the genetic algorithm [119, 201] to infer gene regulatory networks. Another model is the state space model [22, 173, 222], whose main feature is that the gene expression value depends not only on the current internal state variables but also on the external inputs. It is very interesting that the external input is viewed as the previous time step observation, and the gene regulation matrix is obtained from the relationship between the current measurement, the previous measurement, and internal state variables [22, 173]. For the use of state space models, the measurements need to be accurate and a suitable dimension for the internal state variables needs to be determined beforehand, which raises considerable difficulties in experimentation and computation.

In this chapter, we view the gene regulatory network as a dynamic stochastic model, which is composed of the gene measurement equation and the gene regulation equation. In order to reflect the reality, we consider the gene measurement from microarray as noisy, and assume that the gene regulation equation is an one-order autoregressive (AR) stochastic dynamic process. Note that it is very important to regard the models as stochastic, since the gene expression is of inherent stochasticity. Stochastic models can help conduct more realistic simulations of biological systems, and also set up a practical criterion for measuring the robustness of the mathematical models against stochastic noises. After specifying the model structure, we apply, for the first time, the EM algorithm for identifying both the model parameters and the actual value of gene expression levels. Note that EM algorithm is a learning algorithm that can handle sparse parameter identification and noisy data very well. It is also shown that the EM algorithm can cope with the microarray gene expression data with large number of variables but a small number of observations. Four real-world gene expression data sets are employed to demonstrate the effectiveness of our algorithm, and some indices are used to evaluate the models of inferred gene regulatory networks from the viewpoint of bioinformatics.

The remainder of this chapter is organized as follows. In Section 10.2, a stochastic dynamic model is described for genetic regulatory network, which takes into account the noisy measurement as well as the inherently stochastic phenomenon of the genetic regulatory process. The EM algorithm is introduced in Section 10.3 for handling the sparse parameter identification problem and the noisy data analysis. In Section 10.4, our developed algorithm is applied to four real-world gene expression data sets, and the biological significance is discussed in terms of certain criteria. Further discussion is made in Section 10.5 to explain the advantages and shortcomings of our method. Some concluding remarks and future research topics are provided in Section 10.6.

10.2 Stochastic dynamic model for gene expression data

Measuring gene expression levels by DNA Microarray technologies has made a great progress in understanding the interaction amongst genes and extracting functional information. However, the gene expression data measured are often contaminated by measurement noises in a discrete-time fashion, because gene expression time series represent discrete "snapshots" of gene expression at various time points. Therefore, gene expression levels measured can be modeled as

$$y_i(k) = x_i(k) + v_i(k), \ i = 1, 2, \cdots, n, \ k = 1, 2, \cdots, m, \qquad (10.1)$$

where $y_i(k)$ is the measurement data of the ith gene expression levels from Microarray at time k, $x_i(k)$ is the ith actual gene expression levels which stand for mRNA concentrations and/or protein concentrations at time k, $v_i(k)$ is the measurement noise, n is the number of genes, and m is the measurement time points. Without loss of generality, we assume that $v_i(k)$ is a zero mean Gaussian white noise sequence with covariance $V_i > 0$.

Next, we model the gene regulatory network containing n genes by the following stochastic discrete-time dynamic system:

$$x_i(k+1) = -\lambda_i x_i(k) + \sum_{j=1}^{n} a_{i,j} x_j(k) + w_i(k) \qquad (10.2)$$

for $i = 1, 2, \cdots, n$ and $k = 1, 2, \cdots, m$, where λ_i is the self-degradation rate of the ith gene expression product, $a_{i,j}$ represents the regulatory relationship and degree amongst genes. A positive value for $a_{i,j}$ means the jth gene stimulating the expression of the ith gene and, similarly, a negative value for $a_{i,j}$ stands for the jth gene repressing the expression of the ith gene, while a value of zero indicates that jth gene does not influence the transcription of ith gene. This way, each gene in the organism can have multiple inputs, both positive and negative, of differing strength. $w_i(k)$ is the system noise. We also assume that $w_i(k)$ is a zero mean Gaussian white noise sequence with covariance $W_i > 0$, and $w_i(k)$ and $v_i(k)$ are mutually independent.

REMARK 10.1   Compared to the existing state space models [22, 173, 222], our proposed model (10.1)-(10.2) has three advantages. First, it is more suitable for gene regulatory networks, because the gene-to-gene relationships and interactions are considered, and the stochastic property of genes is taken into account. Second, the discrete-time equations (10.1) and (10.2) are convenient for digital computation since gene expression data from Microarray are actually obtained as a series of discrete-time points by using the present experimental technologies. Finally, since the measurements of mRNA concentration using microarray are inevitably noisy, the noise is included in (10.1) to reflect such a fact.   ∎

REMARK 10.2   The measurement noise $v_i(k)$ and the system noise $w_i(k)$ in the model (10.1)-(10.2) are assumed to be zero mean Gaussian white noises. Such an assumption, however, does not lose the generality. The noises can also be modeled as the colored noises, which does not cause difficulties in our algorithm proposed later. More details about whitening noises can be found in [6]. For simplicity, we only consider the case in which $v_i(k)$ and $w_i(k)$ are zero mean Gaussian white noises.   ∎

Now, denote

$$x(k) = \left[\, x_1(k)\; x_2(k)\; \cdots\; x_n(k)\,\right]^T \tag{10.3}$$

for $k = 1, 2, \cdots, m$, and

$$a_i = \left[\, a_{i,1}\; a_{i,2}\; \cdots\; -\lambda_i + a_{i,i}\; \cdots\; a_{i,n}\,\right] \tag{10.4}$$

for $i = 1, 2, \cdots, n$, respectively. We can rewrite (10.2) as follows:

$$x_i(k+1) = a_i x(k) + w_i(k), \quad i = 1, 2, \cdots, n. \tag{10.5}$$

In this chapter, our aim is to establish the model (10.1) and (10.5) from the measurement data

$$
\begin{aligned}
Y := \{ & y_1(1),\; y_1(2),\; \cdots,\; y_1(m), \\
& y_2(1),\; y_2(2),\; \cdots,\; y_2(m),\; \cdots, \\
& y_n(1),\; y_n(2),\; \cdots,\; y_n(m) \}.
\end{aligned}
$$

This is a system identification problem. Notice that for gene expression time series data we typically have a *large* number of variables but a *small* number of observations. Unfortunately, traditional identification methods, such as the least square method, cannot be suitably used for the system identification problem with large number of variables but small number of observations, since they basically require a large amount of observations. Therefore, this problem becomes an "underdetermined" one from the viewpoint of system identification. To handle the data shortage problem, we introduce the expectation-maximization (EM) algorithm to identify the model (10.1) and (10.5). Before introducing our algorithm, we define the vector

$$\theta = \begin{bmatrix} \bar{a} & \bar{W} & \bar{V} \end{bmatrix}, \qquad (10.6)$$

where

$$\bar{a} := \begin{bmatrix} a_1 & a_2 & \cdots & a_n \end{bmatrix},$$
$$\bar{W} := \begin{bmatrix} W_1 & W_2 & \cdots & W_n \end{bmatrix},$$
$$\bar{V} := \begin{bmatrix} V_1 & V_2 & \cdots & V_n \end{bmatrix},$$

which consists of all parameters to be estimated in (10.1) and (10.5).

In the next section, we will develop a computationally efficient iterative method based on EM algorithm for identifying the parameter θ.

10.3 An EM algorithm for parameter identification

In this section, we first introduce the main idea of the expectation-maximization (EM) algorithm. Then, to solve the specified gene network modeling problem, we will derive the iterative computation procedure for the proposed model (10.1) and (10.5) by using the Kalman filtering and Kalman smoothing approaches.

The EM algorithm for time series analysis was first presented by Shumway and Stoffer [188]. It is a general iterative method to compute maximum likelihood (ML) estimates of a set of parameters. It has been shown in various signal processing applications that the use of EM algorithm leads to computationally efficient estimation algorithms [220, 231].

The EM algorithm can be divided into two steps: E-step and M-step. The E-step is to estimate the logarithm likelihood of the complete data using the observed data and the current parameter estimate, and the M-step is to maximize the estimated logarithm likelihood function to obtain the new parameter estimate. Here, given the observations Y and the current parameter estimate, we define the natural logarithm of the conditional expectation of probability density functions for the completed data as the following logarithm likelihood function [188]:

$$J(\theta, \theta^{(l)}) = \mathbb{E}_{\theta^{(l)}} \left[L(X, Y, \theta) | Y \right] \qquad (10.7)$$

where

$$L(X, Y, \theta)$$

$$= -\sum_{i=1}^{n} \left\{ \frac{m}{2} \ln |W_i| + \frac{1}{2} \sum_{k=1}^{m} [x_i(k) - a_i x(k-1)]^T W_i^{-1} \right.$$

$$\times [x_i(k) - a_i x(k-1)] + \frac{(m+1)}{2} \ln |V_i|$$

$$\left. + \frac{1}{2} \sum_{k=0}^{m} [y_i(k) - x_i(k)]^T V_i^{-1} [y_i(k) - x_i(k)] \right\} + C, \qquad (10.8)$$

with the constant C being independent of θ.

The new parameter estimate can be obtained by

$$\theta^{(l+1)} = \arg \max_{\theta} J(\theta, \theta^{(l)}). \qquad (10.9)$$

Next, the new parameter estimate $\theta^{(l+1)}$ can be found by maximizing $J(\theta, \theta^{(l)})$. We maximize $J(\theta, \theta^{(l)})$ with respect to a_i, W_i^{-1} and V_i^{-1}, respectively, and obtain

$$\frac{\partial J}{\partial a_i} = \mathbb{E}_{\theta^{(l)}} \left\{ \sum_{k=1}^{m} W_i^{-1} [x_i(k) - a_i x(k-1)] x(k-1)^T | Y \right\}$$

$$= 0, \qquad (10.10)$$

$$\frac{\partial J}{\partial W_i^{-1}} = \mathbb{E}_{\theta^{(l)}} \left\{ \frac{m}{2} W_i - \frac{1}{2} \sum_{k=1}^{m} [x_i(k) - a_i x(k-1)] \right.$$

$$\left. \times [x_i(k) - a_i x(k-1)]^T | Y \right\} = 0, \qquad (10.11)$$

$$\frac{\partial J}{\partial V_i^{-1}} = \mathbb{E}_{\theta^{(l)}} \left\{ \frac{m+1}{2} V_i - \frac{1}{2} \sum_{k=0}^{m} [y_i(k) - x_i(k)] \right.$$

$$\left. \times [y_i(k) - x_i(k)]^T | Y \right\} = 0. \qquad (10.12)$$

From (10.10)-(10.12), we have

$$a_i^{(l+1)} = \left\{ \sum_{k=1}^{m} \mathbb{E}_{\theta^{(l)}} [x_i(k) x(k-1)^T | Y] \right\}$$

$$\times \left\{ \sum_{k=1}^{m} \mathbb{E}_{\theta^{(l)}} [x(k-1) x(k-1)^T | Y] \right\}^{-1}, \qquad (10.13)$$

$$W_i^{(l+1)} = \frac{1}{m} \left\{ \sum_{k=1}^{m} \mathbb{E}_{\theta^{(l)}}[x_i(k)x_i^T(k)|Y] \right.$$

$$-a_i^{(l+1)} \sum_{k=1}^{m} \mathbb{E}_{\theta^{(l)}}[x(k-1)x_i^T(k)|Y]$$

$$-\sum_{k=1}^{m} \mathbb{E}_{\theta^{(l)}}[x_i(k)x^T(k-1)|Y](a_i^{(l+1)})^T$$

$$+a_i^{(l+1)} \sum_{k=1}^{m} \mathbb{E}_{\theta^{(l)}}[x(k-1)x^T(k-1)|Y]$$

$$\left. \times (a_i^{(l+1)})^T \right\}, \tag{10.14}$$

$$V_i^{(l+1)} = \frac{1}{m+1} \left\{ \sum_{k=0}^{m} [y_i(k)y_i^T(k)] \right.$$

$$-\sum_{k=0}^{m} \mathbb{E}_{\theta^{(l)}}[x_i(k)|Y]y_i^T(k)$$

$$-\sum_{k=0}^{m} y_i(k)\mathbb{E}_{\theta^{(l)}}[x_i^T(k)|Y]$$

$$\left. +\sum_{k=0}^{m} \mathbb{E}_{\theta^{(l)}}[x_i(k)x_i^T(k)|Y] \right\}, \tag{10.15}$$

The EM algorithm is an iterative numerical method for computing the maximum likehood estimate. Letting θ^0 be the initial parameter estimate, the EM algorithm generates a sequence of parameter estimates as follows:

1. *E-step*
Set $\theta = \theta^{(l)}$ and compute $J(\theta, \theta^{(l)})$ in (10.7).
2. *M-step*
Compute $a_i^{(l+1)}$, $W_i^{(l+1)}$ and $V_i^{(l+1)}$ in (10.13)-(10.15) from $i = 1$ to n.

Obviously, in order to compute (10.7) and (10.13)-(10.15), we should first get the conditional expectations for $\mathbb{E}_{\theta^{(l)}}[x_i(k)|Y]$, $\mathbb{E}_{\theta^{(l)}}[x_i(k)x_i^T(k)|Y]$, and $\mathbb{E}_{\theta^{(l)}}[x_i(k)x(k-1)^T|Y]$. In the following, we will provide the Kalman filtering and Kalman smoothing algorithms to compute them.

Before giving the algorithm, we denote

$$\hat{x}^{(l)}(k|m) := \mathbb{E}_{\theta^{(l)}}[x(k)|Y], \tag{10.16}$$

$$\Sigma^{(l)}(k|m) := \mathbb{E}_{\theta^{(l)}}\left\{ [x(k) - \hat{x}^{(l)}(k|m)] \right.$$

$$\left. \times [x(k) - \hat{x}^{(l)}(k|m)]^T |Y \right\}, \tag{10.17}$$

$$\Pi^{(l)}(k, k-1|m) := \mathbb{E}_{\theta^{(l)}}\left\{ [x(k) - \hat{x}^{(l)}(k|m)][x(k-1) \right.$$

$$\left. -\hat{x}^{(l)}(k-1|m)]^T |Y \right\}. \tag{10.18}$$

Since

$$\mathbb{E}_{\theta^{(l)}}\left\{ [x(k) - \hat{x}^{(l)}(k|m)][x(k) - \hat{x}^{(l)}(k|m)]^T |Y \right\}$$

$$= \mathbb{E}_{\theta^{(l)}}[x(k)x^T(k)|Y] - \hat{x}^{(l)}(k|m)[\hat{x}^{(l)}(k|m)]^T,$$

and

$$\mathbb{E}_{\theta^{(l)}}\left\{ [x(k) - \hat{x}^{(l)}(k|m)][x(k-1) - \hat{x}^{(l)}(k-1|m)]^T |Y \right\}$$

$$= \mathbb{E}_{\theta^{(l)}}[x(k)x^T(k-1)|Y] - \hat{x}^{(l)}(k|m)[\hat{x}^{(l)}(k-1|m)]^T,$$

we can obtain

$$\mathbb{E}_{\theta^{(l)}}[x(k)x^T(k)|Y] \quad \text{and} \quad \mathbb{E}_{\theta^{(l)}}[x(k)x^T(k-1)|Y]$$

from $\hat{x}^{(l)}(k|m)$, $\Sigma^{(l)}(k|m)$ and $\Pi^{(l)}(k, k-1|m)$.

The computation of the conditional expectations in (10.16)-(10.18) can be carried out using the Kalman filtering and smoothing methods. To do that, we may represent (10.1) and (10.5) in the following state-space form:

$$x(k+1) = Ax(k) + w(k), \tag{10.19}$$

$$y(k) = x(k) + v(k), \tag{10.20}$$

where

$$A = \begin{bmatrix} a_1 \\ a_2 \\ \vdots \\ a_n \end{bmatrix}, \quad w(k) = \begin{bmatrix} w_1(k) \\ w_2(k) \\ \vdots \\ w_n(k) \end{bmatrix},$$

$$y(k) = \begin{bmatrix} y_1(k) \\ y_2(k) \\ \vdots \\ y_n(k) \end{bmatrix}, \quad v(k) = \begin{bmatrix} v_1(k) \\ v_2(k) \\ \vdots \\ v_n(k) \end{bmatrix}. \tag{10.21}$$

Denoting the current parameter estimate $\theta = \theta^{(l)}$, then A, Q and R is replaced by $A^{(l)}$, $Q^{(l)}$ and $R^{(l)}$, where

$$Q = \text{diag}\{W_1, W_2, \cdots, W_n\}, \quad R = \text{diag}\{V_1, V_2, \cdots, V_n\}.$$

Therefore, at the current iteration cycle, $\hat{x}^{(l)}(k|m)$, $\Sigma^{(l)}(k|m)$ and $\Pi^{(l)}(k, k-1|m)$ can be obtained from the following algorithm [95, 231]:

Forward (filtering) recursions

for $k = 1, 2, \cdots, m$

Propagation Equations

$$\hat{x}^{(l)}(k+1|k) = A^{(l)}\hat{x}^{(l)}(k|k) \tag{10.22}$$

$$\Sigma^{(l)}(k|k-1) = A^{(l)}\Sigma^{(l)}(k-1|k-1)(A^{(l)})^T$$
$$+Q^{(l)} \tag{10.23}$$

Updating Equations

$$K^{(l)}(k) = \Sigma^{(l)}(k|k-1)[\Sigma^{(l)}(k|k-1) + R^{(l)}]^{-1} \tag{10.24}$$

$$\hat{x}^{(l)}(k|k) = \hat{x}^{(l)}(k|k-1) + K^{(l)}(k)[y(k)$$
$$-C\hat{x}^{(l)}(k|k-1)] \tag{10.25}$$

$$\Sigma^{(l)}(k|k) = \Sigma^{(l)}(k|k-1) - \Sigma^{(l)}(k|k-1)[\Sigma^{(l)}(k|k-1)$$
$$+R^{(l)}]^{-1}\Sigma^{(l)}(k|k-1) \tag{10.26}$$

Backward (smoothing) recursions

for $k = m, m-1, \cdots, 1$

$$\Lambda^{(l)}(k-1) = \Sigma^{(l)}(k-1|k-1)(A^{(l)})^T$$
$$\times[\Sigma^{(l)}(k|k-1)]^{-1} \tag{10.27}$$

$$\hat{x}^{(l)}(k-1|N) = \hat{x}^{(l)}(k-1|k-1) + \Lambda^{(l)}(k-1)$$
$$\times[\hat{x}^{(l)}(k|N) - \hat{x}^{(l)}(k|k-1)] \tag{10.28}$$

$$\Sigma^{(l)}(k-1|N) = \Sigma^{(l)}(k-1|k-1) + \Lambda^{(l)}(k-1)$$
$$\times[\Sigma^{(l)}(k|N) - \Sigma^{(l)}(k|k-1)]$$
$$\times\Lambda^{(l)}(k-1)^T \tag{10.29}$$

and

$$\Pi^{(l)}(k, k-1|m) = \Pi^{(l)}(k, k-1|k) + [\Sigma^{(l)}(k|m)$$
$$- \Sigma^{(l)}(k|k)][\Sigma^{(l)}(k|k)]^{-1}$$
$$\times \Pi^{(l)}(k, k-1|k) \tag{10.30}$$

$$\Pi^{(l)}(k, k-1|k) = [I - K_k^{(l)}]A^{(l)}\Sigma^{(l)}(k-1|k-1) \tag{10.31}$$

REMARK 10.3 The EM algorithm is only guaranteed to converge to a local maximum of the likehood function. Therefore, in order to ensure convergence to the global maximum, a good initialization procedure may be required.

To initialize the Kalman smoothing equations, we need to specify $\hat{x}^{(l)}(0|0)$ and $\Sigma^{(l)}(0|0)$. We may use the first observed data to specify $\hat{x}^{(0)}(0|0)$ and $\Sigma^{(0)}(0|0)$. These initial estimates can then be iteratively improved by using the final estimates from the previous iteration cycle, i.e., $\hat{x}^{(l+1)}(0|0) = \hat{x}^{(l)}(0|m)$ and $\Sigma^{(l+1)}(0|0) = \Sigma^{(l)}(0|m)$. ∎

REMARK 10.4   Since biologically the resulting gene regulatory is expected to be sparse, we set some of the matrix entries equal to zero, and infer the network using only the nonzero entries. If we know some parameters $a_{i,j}$ *a priori*, we don't need to include the known $a_{i,j}$ in θ for computation and only specify them in the matrix A. Moreover, if one group of genes are not related with other groups of genes, we can divide them into several groups. Several small gene regulatory networks are only computed, which reduces the computational complexity. Note that other conventional system identification algorithms, such as least square method, cannot be used to deal with the sparse data in such an effective way. ∎

10.4   Simulation Results

In order to evaluate the performances of the proposed algorithm, we adopt four real-world gene expression data sets, that is, the yeast gene expression time series [228], the virus gene expression time series [119], the human malaria gene expression time series [31], and the worm gene expression time series [21,137]. Our modeling process is carried out after data pre-processing. Normalization is applied to the gene expression profile by taking log ratios first and then mean centering. After the normalization, the aforementioned EM algorithm is employed to the these data sets, in order to model the gene expression network dynamics.

In order to be concise, we will elaborate the identification process for yeast and virus gene expression time series in subsections 10.4.1 and 10.4.2, but will briefly describe the simulation results for dynamic modeling of human malaria and worm gene expression time series in Subsection 10.4.3.

10.4.1   Modeling of yeast gene expression time series

The first data set is from the yeast gene expression experiment, which consists of expression levels of 237 genes at 17 equally-spaced time points, selected by Yeung et al. [228]. This data set is available from the website http://faculty.washington.edu/kayee/model/.

By using the proposed EM algorithm, for the first data set, the gene regulation matrix for group 3 of yeast gene expression experiment is obtained by

$$A = \begin{bmatrix} A_1 & A_2 & A_3 & A_4 \end{bmatrix},$$

where

$$A_1 = \begin{bmatrix}
-0.4824 & -0.3440 & -0.4248 & 0.5235 \\
-0.0579 & -0.0962 & 0.1152 & 0.3520 \\
-0.1092 & -0.1901 & 0.2481 & 0.2095 \\
0.0161 & 0.3041 & -0.1126 & -0.0678 \\
-0.1161 & 1.0210 & 0.5017 & 0.0996 \\
0.2056 & -0.0771 & -0.0145 & 0.2066 \\
0.2625 & 0.5455 & -0.2064 & -0.1530 \\
0.1012 & 0.0835 & 0.1955 & 0.0490 \\
0.1452 & 0.6964 & -0.0621 & 0.2799 \\
-0.0162 & 0.2625 & -0.3951 & -0.3521 \\
0.2221 & 0.2684 & 0.0338 & -0.0121 \\
0.2114 & -1.2157 & 0.3920 & -0.0611 \\
0.0352 & 0.2100 & -0.2358 & -0.1400 \\
0.2967 & 0.3958 & -0.1568 & 0.4209 \\
-0.2405 & 0.3095 & 0.0738 & 0.2097 \\
0.1657 & -0.1242 & -0.1044 & 0.1269 \\
0.3554 & -0.2441 & 0.0842 & -0.2436 \\
-0.0808 & 0.2869 & 0.1084 & 0.2325
\end{bmatrix},$$

$$A_2 = \begin{bmatrix}
0.3822 & -0.4154 & 0.6976 & 0.3902 \\
-0.0779 & -0.1062 & 0.0411 & -0.1198 \\
-0.0055 & 0.1430 & -0.0605 & -0.1433 \\
-0.1018 & -0.2795 & -0.1204 & -0.1456 \\
-0.0030 & 0.8956 & -0.0261 & -0.2367 \\
-0.0373 & -0.1362 & 0.1804 & -0.1333 \\
0.1482 & -0.5774 & -0.0740 & -0.5611 \\
0.1986 & 0.5117 & -0.0467 & -0.3921 \\
0.0441 & 0.3137 & 0.2650 & -0.6973 \\
-0.2136 & -0.2032 & -0.2038 & -0.3563 \\
0.2217 & -0.5210 & -0.1232 & 0.1624 \\
0.3203 & 0.3511 & -0.0671 & 0.8208 \\
-0.2304 & 0.0839 & -0.0543 & 0.0862 \\
0.0581 & 0.0183 & 0.1491 & 0.0122 \\
0.0047 & -0.0408 & -0.0051 & -0.2914 \\
-0.1694 & 0.2038 & 0.2998 & 0.0877 \\
0.3571 & 0.4504 & 0.1233 & 0.2794 \\
-0.0934 & -0.0234 & -0.0159 & -0.0276
\end{bmatrix},$$

$$A_3 = \begin{bmatrix} 0.5201 & -0.5319 & 0.1039 & -0.2079 & -0.4381 \\ 0.6151 & 0.2917 & -0.2760 & 0.4203 & 0.0156 \\ 0.1162 & 0.4696 & 0.1618 & 0.0102 & 0.0221 \\ 0.1277 & 0.3725 & -0.2130 & -0.2519 & 0.3515 \\ -0.7025 & -0.0071 & 0.0810 & -0.4779 & 0.3996 \\ 0.3043 & 0.2478 & 0.2310 & -0.4495 & -0.0246 \\ 0.0397 & -0.1946 & 0.2223 & -0.5854 & 0.0124 \\ 0.0190 & 0.1391 & 0.5357 & 0.2563 & 0.1859 \\ 0.2149 & 0.1320 & 0.1059 & -0.1450 & 0.1126 \\ 0.1770 & -0.4520 & -0.4880 & 0.1747 & 0.5669 \\ 1.1356 & -0.3625 & 0.1256 & 0.5414 & 0.6818 \\ -0.3664 & -0.0691 & -0.3463 & 0.0199 & -0.1555 \\ -0.0724 & 0.0897 & 0.0350 & 0.0254 & 0.5366 \\ -0.3639 & -0.2656 & 0.3253 & 0.1790 & 0.4707 \\ 0.6284 & 0.5105 & 0.0863 & -0.0651 & 0.1140 \\ 0.3486 & 0.3826 & 0.2239 & -0.1961 & -0.1705 \\ -0.0905 & 0.6867 & -0.3379 & 0.0683 & 0.0780 \\ -0.0028 & 0.2788 & -0.1808 & 0.0664 & -0.0564 \end{bmatrix},$$

$$A_4 = \begin{bmatrix} -0.7611 & 0.1691 & -0.1328 & -0.4932 & 0.1544 \\ 0.0058 & 0.1407 & -0.1606 & -0.2516 & 0.7453 \\ -0.2250 & 0.4366 & -0.0785 & -0.2756 & 0.0278 \\ -0.1879 & 0.1373 & 0.1641 & 0.2014 & -0.1100 \\ 0.7104 & -0.1843 & 0.1939 & -0.1773 & -1.3029 \\ -0.2462 & 0.2970 & -0.0906 & -0.2234 & -0.4861 \\ 0.2977 & -0.0892 & -0.0805 & 0.3771 & 0.5676 \\ -0.2419 & 0.0550 & 0.0537 & -0.2569 & 0.4901 \\ 0.1182 & 0.0445 & 0.0969 & -0.2355 & -1.0189 \\ -0.4872 & -0.5928 & 0.2526 & -0.0132 & 0.7997 \\ -0.2117 & -0.1965 & -0.0988 & -0.1643 & -0.0749 \\ 0.2701 & -0.4681 & -0.3162 & 0.3189 & 0.7409 \\ -0.4739 & -0.0497 & 0.4096 & 0.0768 & 0.2256 \\ 0.1121 & 0.1047 & 0.1813 & 0.2154 & -0.7320 \\ 0.0285 & 0.0575 & -0.0551 & -0.0846 & -0.5759 \\ -0.3424 & -0.0993 & 0.0050 & -0.1188 & 0.1565 \\ -0.1529 & -0.3316 & -0.0158 & -0.1467 & 0.1716 \\ -0.1127 & 0.3264 & 0.0387 & 0.1280 & 0.0218 \end{bmatrix}.$$

The covariance of the yeast gene network model is calculated as

$$W = \begin{bmatrix} 0.0555 & 0.0075 & 0.0109 & 0.0083 & 0.0388 & 0.0126 \\ 0.0365 & 0.0076 & 0.0051 & 0.0114 & 0.0073 & 0.0253 \\ 0.0090 & 0.0090 & 0.0132 & 0.0203 & 0.0231 & 0.0020 \end{bmatrix},$$

and the covariance of the yeast gene expression measurement noise is computed by

$$V = \begin{bmatrix} 0.1327 \; 0.0006 \; 0.0311 \; 0.0105 \; 0.0787 \; 0.0011 \\ 0.1682 \; 0.0075 \; 0.0170 \; 0.0021 \; 0.0017 \; 0.0157 \\ 0.0012 \; 0.0014 \; 0.0706 \; 0.0805 \; 0.0661 \; 0.0001 \end{bmatrix}.$$

As we can see from above, all the parameters of the proposed stochastic dynamic model can be easily obtained by using our algorithm. Furthermore, the predicted values of gene expression levels are also obtained. We can observe that, for the gene expression levels, there exist differences between the actual values and the predicted (simulated) values, and the prediction errors of every yeast gene are shown in Figs. 10.1-10.4.

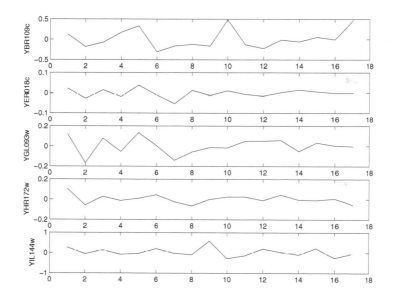

FIGURE 10.1

The measurement errors of yeast genes (part 1)

10.4.2 Modeling of virus gene expression time series

The second data set is for the virus gene expression microarray data from [119], which consists of 106 genes expressed at 8 equally-spaced time points.

Again, by using the proposed EM algorithm to the second data set, the gene regulation matrix for group one of virus gene expression experiment is obtained by

$$A = [A_1 \; A_2]$$

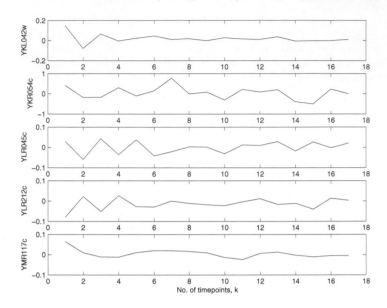

FIGURE 10.2

The measurement errors of yeast genes (part 2)

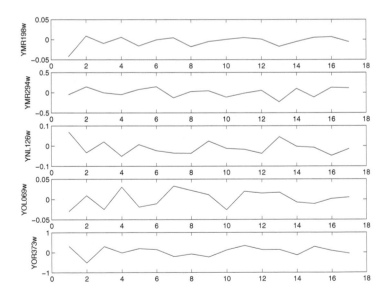

FIGURE 10.3

The measurement errors of yeast genes (part 3)

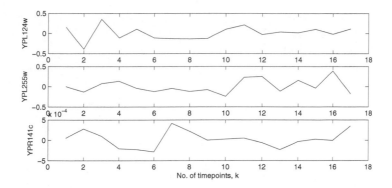

FIGURE 10.4
The measurement errors of yeast genes (part 4)

where

$$A_1 = \begin{bmatrix} -0.0380 & 0.2864 & 0.0224 & 0.1894 & -0.1095 \\ 0.0358 & -0.0149 & 0.4452 & 0.2760 & -0.3539 \\ -0.0578 & 0.2458 & 0.2424 & 0.2885 & -0.2094 \\ 0.5355 & -0.0647 & 0.7688 & 0.1994 & -0.0052 \\ 0.0727 & 0.4132 & 0.1664 & 0.0991 & -0.1088 \\ 0.1106 & 0.1621 & 0.2381 & 0.1420 & 0.0652 \\ -0.0108 & 0.2278 & 0.2461 & 0.1063 & -0.1258 \\ 0.1280 & 0.1246 & 0.2874 & 0.0801 & 0.0095 \\ -0.0758 & 0.2457 & 0.1075 & -0.1461 & -0.2045 \\ 0.3163 & -0.3176 & 0.0884 & -0.0962 & -0.2591 \end{bmatrix}$$

$$A_2 = \begin{bmatrix} 0.0563 & 0.1044 & 0.2009 & 0.4790 & 0.2404 \\ 0.0146 & 0.0201 & -0.0330 & -1.1245 & 0.4650 \\ 0.0932 & 0.1267 & 0.1010 & -0.0704 & 0.0864 \\ 0.0681 & 0.2249 & -0.0024 & 0.6778 & 0.3909 \\ -0.0145 & -0.0234 & -0.0024 & -0.1631 & 0.3268 \\ -0.0501 & 0.0183 & -0.0303 & 0.2039 & 0.2081 \\ 0.0094 & -0.0296 & -0.0394 & 0.1782 & 0.0468 \\ -0.0049 & 0.0025 & -0.1375 & 0.4621 & 0.1838 \\ -0.0659 & 0.0181 & -0.1992 & -0.1900 & -0.2829 \\ -0.0060 & 0.2219 & 0.1804 & 0.8216 & -0.0740 \end{bmatrix}$$

The covariance of viral gene network model is

$$W = \begin{bmatrix} 0.0900 & 0.0560 & 0.0803 & 0.2862 & 0.0688 \\ 0.1253 & 0.1194 & 0.0753 & 0.0263 & 0.1453 \end{bmatrix}$$

and the covariance of viral gene expression measurement noise is

$$V = \begin{bmatrix} 1.0196 & 0.0350 & 0.4137 & 9.5014 & 0.5865 \\ 4.2930 & 4.7446 & 0.9643 & 0.0000 & 11.4422 \end{bmatrix}$$

All the parameters of stochastic dynamic model as well as the noise intensity are simultaneously calculated, and the prediction errors of every virus gene are illustrated in Figs. 10.5-10.6.

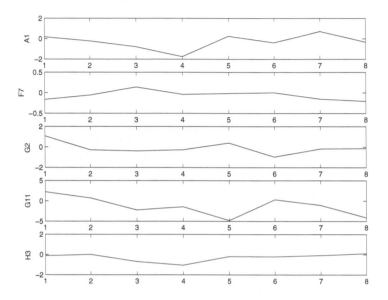

FIGURE 10.5

The measurement errors of virus genes (part 1)

10.4.3 Modeling of human malaria and worm gene expression time series

The third data set is from the human malaria gene expression time series [31]. As stated in [31], *Plasmodium falciparum* is responsible for the vast majority of the 300-500 million episodes of malaria worldwide and accounts for 0.7-2.7 million annual deaths. A comprehensive understanding of *Plasmodium* molecular biology will be essential for the development of new chemotherapeutic and vaccine strategies. Therefore, it is of great importance to model the human malaria expression data, which are made throughout the invasion process, with no observable abrupt change in the expression program upon successful reinvasion. The human malaria expression data set consists of 530 genes expressed at 48 equally-spaced time points. We select a group of 15 genes and apply the proposed EM algorithm. All the model parameters can be obtained, which are not given here for the purpose of saving space. To illustrate the usefulness of the proposed modeling method, we display the prediction errors of

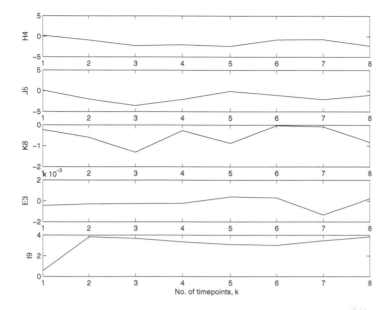

FIGURE 10.6
The measurement errors of virus genes (part 2)

every human malaria gene in Figures 10.7-10.9.

The fourth data set is from the worm gene expression time series [21, 137], which consists of 98 genes expressed at 123 equally-spaced time points. Again, we select a group of 15 genes, apply the proposed EM algorithm Figures' and display the prediction errors of the selected worm gene in Figures 10.10-10.12.

10.5 Discussions

10.5.1 Model quality evaluation

Since it is generally difficult to understand the real gene regulatory networks completely by biological experiments at present, some researchers [221, 222] proposed several indices to evaluate the models for gene regulatory networks from the viewpoint of bioinformatics, such as the computational cost, the prediction power (error), the stability, the robustness, and the periodicity. Obviously, different evaluation standards should be applied to different kinds of models. Since our models are stochastic and the measurements are noisy, our evaluation indices will mainly focus on the computational cost, the estimation covariance, the stability, and the robustness.

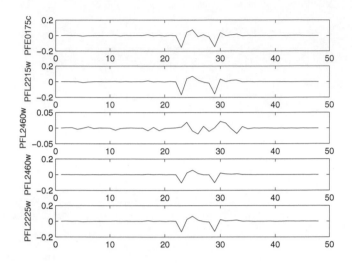

FIGURE 10.7

The measurement errors of human malaria genes (part 1)

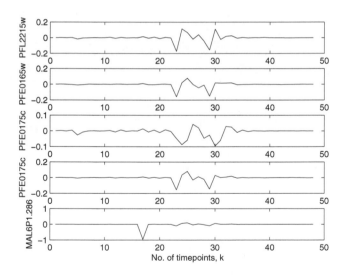

FIGURE 10.8

The measurement errors of human malaria genes (part 2)

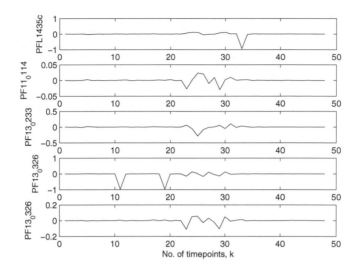

FIGURE 10.9

The measurement errors of human malaria genes (part 3)

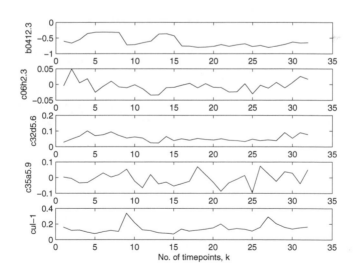

FIGURE 10.10

The measurement errors of worm genes (part 1)

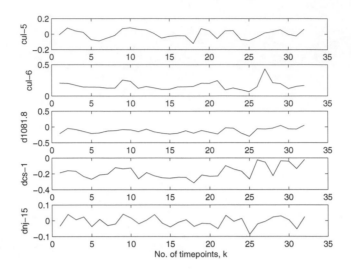

FIGURE 10.11

The measurement errors of worm genes (part 2)

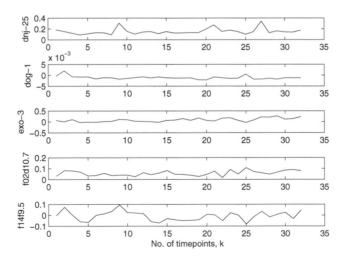

FIGURE 10.12

The measurement errors of worm genes (part 3)

GETS	Yeast	Virus	Malaria	Worm
MER	10.60%	42.19%	23.26%	28.91%

FIGURE 10.13

Quantitative model evaluation

For the computational cost, our EM algorithm is an iterative learning algorithm that doesn't involve searching. The computational complexity is only related with the number of genes, the time points and the number of iteration. From our simulations on the four data sets, the computational time is in seconds on a PC computer, hence the computational cost is light.

From our experiments on the four data sets, the estimation covariances W are small, which means that our models fit the data very well. The covariances V represent the quality of measuring gene expression levels using microarray. For example, examining the covariances V for the established yeast and virus time series models, we can see that the measurement of yeast gene expression levels is accurate whereas the measurement of virus gene expression levels is not quite accurate, because the covariances V is smaller for the yeast measurement and the covariances V is bigger for the virus measurement. Furthermore, in order to evaluate the model quality in a quantitative way, let us introduce the following criterion for the modeling errors (error ratio in percentage) between the actual and predicted data [134]:

$$\text{Error Ratio} = 100 \times \frac{1}{l} \sum_{c=1}^{l} \left[\sqrt{\frac{\sum_{k=1}^{s} (y_{ck} - \hat{y}_{ck})^2}{\sum_{k=1}^{s} y_{ck}^2}} \right] \% \qquad (10.32)$$

where l is the number of genes (dimension) involved in the modeling; s is the number of observations (length); y_{ck} is the actual gene expression value for c-th gene at the k-th time point. The results are given in the following table where GETS stands for the gene expression time series and MER means the modeling error ratio:

It can be seen from Figure 10.13 that the model quality is generally satisfactory. The publicly available yeast gene expression time series data is of a good quality that leads to the best model. It is not surprising that the model for the virus gene expression time series is relatively the worst simply because of the poor quality of the data set (only 8 observations are made for each gene). In fact, the lengths for all the four time series considered here are very short and, as will be discussed later, traditional modeling approaches fail to cope with the short time series modeling due to the assumption on the length of the time series.

In order to check the stability and the robustness of our models, we need to compute the eigenvalues of the regulation matrices of the four models. For the virus gene expression time series, the malaria gene expression time series and the worm gene expression time series, the sets of the eigenvalues of the regulation matrix are, respectively,

$$\mathbb{E}_{\text{virus}} = \{-0.1277 \pm 0.8784i, 0.0893 \pm 0.4355i, 0.8692,$$
$$-0.4490, -0.3501, -0.0729, -0.0604, -0.0611\},$$
$$\mathbb{E}_{\text{malaria}} = \{-0.4902 \pm 0.7282i, -0.8811, 0.3662 \pm 0.6502i,$$
$$0.6410, -0.3408, 0.1044, -0.1660, -0.1055,$$
$$-0.0847, -0.0486 \pm 0.0046i,$$
$$-0.0606 \pm 0.0038i\},$$
$$\mathbb{E}_{\text{worm}} = \{0.9993, 0.5957, -0.1149 \pm 0.2028i, 0.0307,$$
$$-0.0953 \pm 0.0522i, -0.0461, -0.0562,$$
$$-0.0875 \pm 0.0014i, -0.0720, -0.0807,$$
$$-0.0767 \pm 0.0017i\}$$

Obviously, for the models of virus, malaria and worm gene expression time series, all eigenvalues lie well inside the unit circle. Therefore, the models are stable and robust. For the yeast data set, eighteen eigenvalues of the regulation matrix are given by $0.9067 \pm 0.3304i, 0.7305 \pm 0.6659i, 0.3706 \pm 0.8439i, 0.0406 \pm 0.9303i, -0.2429 \pm 0.8529i, -0.5551 \pm 0.6032i, -0.7877 \pm 0.3466i, -0.0063 \pm 0.0102i, -1.0260, -0.3956$. All of these except one lie inside the unit circle. Although one of the eigenvalues is outside the unit circle, it is very close to 1. If the gene regulatory network is periodic, this eigenvalue would not cause the instability. Hence, the model is almost stable and robust.

10.5.2   Comparisons with existing modeling methods

In biology and medicine, it is quite common for the multivariate time series (MTS) data to be rather short, either because of the expense involved in obtaining data, e.g. in high-throughput bioinformatics areas such as microarrays, or due to the practicalities such as patients' treatment period or mortality.

Traditionally, statistical methods have been proposed when modeling MTS, e.g., the Vector Auto-Regressive process, Vector Auto-Regressive Moving Average, Markov Chain Monte Carlo methods, and other non-linear and Bayesian systems [42,60]. In the computing community, many MTS forecasting methods have been proposed using recurrent or time-delay neural networks, evolutionary computation, inductive logic programming and support vector machines, see [42,94] and the references therein.

However, one area where there has been little work is the analysis of a particular type of time series in which the data set consists of a large number of variables but with a small number of observations. Almost all traditional methods for modeling MTS place constraints on the minimum number of time series observations in the dataset; many methods require distribution assumptions to be made regarding the observed time series, e.g. the maximum likelihood method for parameter estimation [42]. For example, a traditional

GETS	Yeast	Virus	Malaria	Worm
Dimension	237	106	530	98
Length	17	8	48	123

FIGURE 10.14

Dimension vs. length

way of modeling MTS data is the Vector Auto-Regressive Process, usually denoted as VAR(P) for a model of order P. The standard statistical methods for fitting a VAR process to a set of data often consist of two steps: order selection and parameter estimation. Order selection is commonly performed through the use of information theory based metrics such as Akaike's Information Criterion. Many of these metrics will impose a restriction on the minimum length of a MTS, N, based on the number of degrees of freedom of the model being estimated: $N > KP + 1$ where K is the number of variables being modeled, and P is the order of the VAR process. For example, for a MTS consisting of 100 variables, to find the most appropriate order of a VAR process with a maximum order of three under consideration, N must be at least 302. This restriction is clearly unacceptable for modeling many short, high-dimensional time series, which are common in biology and medicine.

For the four gene expression data sets considered in this chapter, the number of genes (dimension) and the number of observations (length) are given in Figure 10.14 where GETS stands for the gene expression time series . It can be seen clearly from Figure 10.14 that the gene expression time series is pretty short but with high dimensions, for which the traditional MTS modeling methods are simply impossible to be applied in a satisfactory way.

Recently, the modeling problem for short, high-dimensional time series has begun to receive some research interest. For example, dynamic Bayesian networks have been proposed to model gene expression time series data [119, 146] with the ability of handling noisy/hidden variables. However, dynamic Bayesian networks need more complex algorithms such as the genetic algorithm [119, 201] to infer gene regulatory networks, and the noise intensity cannot be obtained directly. In previous sections, we have provided an algorithm to infer a gene regulatory network in the form of stochastic dynamic models. Using our algorithm, the gene regulation matrix can be extracted from *a very short number* of gene expression time series data. This matrix can be employed to figure out how genes act "in concert" to achieve specific phenotypic characteristics [221].

Compared to existing MTS modeling methods, our algorithm has the following advantages:

(1) Our algorithm can effectively tackle short, high-dimensional time series that typically occurs in biology and medical sciences.

(2) The defined gene regulation matrix can reflect the relationship and in-

teraction between genes, where a_{ij} stands for the effect of jth gene on ith gene. The model is of direct biological significance.

(3) Our scheme can identify the entire network structure. For the existing differential equation method, there exists an "underdetermined" problem if the number of genes is equal to or larger than the number of experiments due to the shortage of gene expression data. In this paper, we use iterative "learning" procedure so that the global dynamic model can be obtained from a small number of gene expression data.

(4) Our method can deal with noises in gene expression data measurement and sparse connectivity. Since the measurements of mRNA concentration using microarrays are typically noisy, it is very advantageous that our algorithm is robust to noises while identifying gene regulatory networks.

(5) Our algorithm can tackle the spare gene regulatory networks only by setting some of the matrix entries as zero. The algorithm is especially efficient for lager gene regulatory networks which can be divided into several individual gene regulatory networks.

(6) Our algorithm is very efficient for computation, and the computational cost is light even if there is a lack of gene expression data.

Nevertheless, our algorithm cannot deal with the data sets with missing values, which often occurs in many gene measurement data. Furthermore, a better data pre-processing method should be explored to maintain the biological meaning while simplifying the computation, since different pre-processing (normalization) method would cause different gene regulation matrices. On the modeling issue, the models would be different for different iteration times and initial conditions. Proper biological knowledge should be incorporated to our algorithm and some constraints need to be added to make the solution set smaller. Moreover, for simplicity we have adopted a stochastic linear model with constant coefficients to infer gene regulatory network. How to consider the influence of time-varying coefficients, nonlinearities and time delays will be the topic of our further investigation.

10.6   Conclusions and Future Work

In this chapter, we have developed an EM algorithm for modeling the gene regulatory networks from gene expression time series data. The gene regulation matrix has been obtained by an iterative learning procedure based on the stochastic linear dynamic model. Gene expression stochastic dynamic models for four real-world gene expression data sets have been constructed to show how our algorithm works. Our algorithm can tackle the spare gene regulatory networks only by setting some of the matrix entries as zero. Furthermore, our algorithm is especially efficient for lager gene regulatory networks which can be divided into several individual gene regulatory networks. Therefore, our algorithm can be ideally applied to modeling the gene regulatory networks where the real connectivity of the network is specified *a priori*. We expect that our algorithm will be useful for reconstructing gene regulatory networks on a genome-wide scale, and hope that our results will benefit for biologists and biological computation scientists. In the near future we will continue to investigate some real-world gene expression data sets and apply our algorithm to reconstruct gene regulatory networks with missing data, sparse connectivity, periodicity and time-delays. We are also getting connection with the biologists to explain our results from the biological point of view.

11

Conclusions

11.1 Introduction

The aim of this chapter is to review the investigation reported in the book. The main motivation for the investigation was to explore the challenges involved with the analysis of microarray image data such that the analysis system is not only enhanced but automated. The need to address this automated issue becomes particularly important as microarray systems offer huge advancements in medical diagnosis and drug discovery processes. However, current approaches to the analysis of microarray images rely a great deal on the human aspect, critically reducing the high throughput capabilities of these systems. Recognizing the importance of the automation issue (along with the related elements of process repeatability and system robustness) particularly, the study focused on delivering a system that encompassed these concepts throughout.

Four areas of investigation were identified as key components of a microarray analysis system and as such important in the creation of an appropriate system. The analysis of these key components produced a set of weaknesses of current analysis systems, be they performance, quality or consistency based. These issues were broadly discussed in Chapter 2 with an example analysis package highlighted to reinforce these issues in an existing and practical sense.

Section 11.2 reviews the achievements in respect to the key issues as highlighted in Chapter 1. Since the study has presented a process that combines the biological and computer science domains, the subsequent Section 11.3 and Section 11.4 highlight the research contributions made by the book in terms of contributions to technical and practical. The final section, Section 11.5, reflects on possible limitations of the study and sets an agenda for future research.

11.2 Achievements

Generally, research in the field of microarray image analysis has relied heavily on the vision system of the human operator. This human intervention has typically been focused on correctly addressing the individual gene spots in the image. The intervention issue has been discussed in detail in Chapter 2 with respect to a microarray image being full of noise artifacts (hair, scratches and dust for example). The existence of such artifacts in an image can lead the underlying addressing algorithms on a proverbial goose chase. As such, current research in the field of microarray image analysis typically proposes hybrid solutions (human optimized algorithms). Such solutions are equipped with some form of detection or identification algorithms, which are further augmented with the results of the human based identification processes.

A good example of such a hybrid solution is that of the GenePix process. Although GenePix does not represent a gold standard, its functionality and implementation processes have been accepted by the microarray biological community at large. The GenePix process requires the operator to manually demarcate all of the genes on a microarray image (with the use of a pre-defined operator created template structure). The GenePix systems addressing algorithms are then able to refine the template position on an individual gene spot basis. Typically, the operator must browse the entire surface of the image again to rectify any algorithm induced position errors. Such browsing wastes more of the operator's time which could be focused on other aspects of research and as such, this browsing process is usually superficial at best.

A more recent system designed for this image analysis task is that of SPOT from the CISO laboratory. In the SPOT process, human intervention has been reduced by requiring the operator to manually identify the upper and lower outer gene spot from all of the meta-blocks in the image. The identification algorithms use this manual position information as the bounding block, and thus search internally for the gene spot members within the given region.

Processes that rely on this manual human addressing task significantly reduce the throughput capabilities of microarray systems. Also, due to the human aspect, the ability to generate consistent result repeatability will be reduced as the operator is unlikely to demarcate a given gene spot in exactly the same way over several days. As such, the manual intervention and poor result repeatability of existing analysis systems were identified as the main contributing factors giving motivation to this study.

The major contribution of the study based on the four key issues can be seen in terms of proposing a framework system that facilitates the fully automatic blind processing of complementary deoxyribonucleic acid (cDNA) microarray image data. The framework's component parts were designed such that they would function without the requirement of any prior image or domain specific knowledge. Indeed, a stringent design requirement from the outset was that the framework processes should be blind to any prior knowledge, with required knowledge learned from the current image as and when needed.

The *Copasetic Microarray Analysis* (CMA) framework has consistently produced accurate analysis results while achieving the aims as set out in Chapter 1. This was further demonstrated with the analysis of a series of relatively "good" and "bad" quality microarray images, whose results have proved to be very promising. Indeed, in many cases, the results are sufficiently good enough for images that were initially thought to contain too much noise to be of practical benefit for biologists.

This performance is all the more successful when it is remembered that this framework operates in a truly blind manner. At the initiation of the framework process, the operator simply passes the raw digitized microarray as the input image to the framework. There is no need for the operator to supply domain specific image knowledge to the system as this is calculated purely from the image data as and when required. However, in cases where the image has extremely significant systematic noise or artifacts, the internal algorithms can become sidetracked. As such, a manual identification stage has been implemented in the appropriate component, but it should be noted that this process is only presented to the operator in cases of extreme situations.

In this way, CMA is able to overcome deficiencies of existing microarray analysis systems. Also, due to the component based construction of the framework, the proposed techniques are generally flexible enough for other domains. For example, there is no reason why the techniques could not be applied to other image analysis areas such as satellite reconnaissance, *magnetic resonance imaging* (MRI) scans, and 2D electrophoresis gels. Indeed, a preliminary investigation examined the framework's performance when given gel images. With a simple modification of structure requirements with respect to the compositional processes of the framework, the initial analysis of the gel images were positive.

The central outcome of the study was a microarray analysis framework comprising of five core tasks. The core tasks emerged as a result of analyzing the weaknesses of existing microarray analysis systems as above, which defined the scope of the study in terms of four general problem areas. The extent to which each of the four problem areas has been met by the CMA framework is discussed in the following subsections.

11.2.1 Noise reduction

Recall that the purpose of the microarray experiment is to help biologists with their understanding of the human biomedical system. This is achieved by measuring the interaction of a microarray's underlying genes with probable disease compounds for example. This means the gene spots in the image must be identified and quantified in some way. Clearly any noise in the image surface will impact on the quantified measurements of the gene spots.

The most critical aspect of a microarray experiment process is that of noise reduction. Without first cleaning the data appropriately, the best analysis process available will be meaningless. The task of the noise reduction processes is obvious, not so clear however is how this reduction can be achieved without introducing some negative effect onto the data. Dimensionality reduction techniques usually sacrifice some form of data accuracy to render the problem space smaller, which in turn means the data can be analyzed more thoroughly.

Packages such as GenePix and SPOT to some extent sidestep this noise identification problem by requiring the operator manually mark the gene spot regions. Template structures are used to simplify the bulk of this identification process, with the operator focusing on the accuracy of the template members primarily. Clearly, any errors introduced at the beginning of the analysis process will propagate through to the final stages.

Also, due to this operator intensive identification and demarcation process the high-throughput potential of the microarray system is not utilized appropriately. As such, the CMA frameworks underlying processes were designed with this noise reduction requirement in mind such that the various stages overlap slightly in functionality to help reduce error propagation and noise. Such an integrated system takes a step in the right direction in releasing the microarrays inherent high-throughput capabilities. Technically speaking, due to the desired overlap in functionality for the CMA processes, the noise reduction aspects of the investigation are relevant to all four core aspects as set out in the aims and objectives discussed here. However, Chapters 3 and 4 deal specifically with different aspects of this reduction concept and as such are briefly mentioned in this section.

Chapter 3 proposed a new approach to microarray image analysis. Rather than rely on the operator to perform various sub-tasks in order to identify the image's gene spots (the penultimate analysis task), Chapter 3 proposed a process that clarifies these interesting feature spaces in an unsupervised manner, such that automated analysis can take place more readily. This feature enhancing application was explored with the design and implementation of the *Image Transformation Engine* (ITE) process. The ITE was created to highlight certain aspects of a typical microarray image as and when required. For example, by giving more weight to pixel intensities in the mid-to-high-end range of the possible 65536 unique values, the ITE process is able to highlight those pixels typically associated with the gene spots. If such an enhancement process is integrated with a rescaling of the 0-65535 values into the new range of 0-255, not only is memory and process time reduced, but knowledge loss (due to rescaling) is minimized.

As already mentioned, there is a designed overlap aspect of the CMA processes that helps with the reduction of error propagation through the framework. Chapter 4 takes advantage of such overlapping ideas and explored a novel method *Pyramidic Contextual Clustering* (PCC) that scales traditional clustering techniques to large scale (twenty million–plus observations) datasets. The PCC process provides the ability to improve on traditional results by utilizing the image's contextual information that is typically discarded by existing techniques. Whereas the ITE process focused on the image's intensity values specifically, PCC is more interested in the spatial distribution of these intensities. At the same time PCC renders transparent the clustering technique's internal decision process. With the correct utilization of this transparency data, it is now possible to harness more knowledge from the clustering processes than would otherwise be possible.

11.2.2 Gene spot identification

With the majority of the noise reduction aspects of microarray image data analysis mentioned above, the focus of investigation becomes one of gene spot identification. A typical microarray image yields approximately 5000×2000 pixels in the digital image representation. Of these twenty million–plus pixel observations, just over 2.2 million represent gene spot member pixels themselves. This means in turn that the remaining pixels represent background noise. However, whereas packages such as GenePix and SPOT rely on the human operator to identify these gene spot regions, an automated system must use some other method.

Chapter 5 explored a technique that determines such microarray compositional information from geometrical principles. The chapter actually proposes two related processes for this compositional determination. The first process, *Image Layout* (IL), searches the microarray image to find inherent structures that have some form of regularity or consistency across their entire region. Such structures are usually associated with the master block positions of the image's gene spots (care must be taken such that misclassification of regular noise is minimized). Once these master block locations have been appropriately determined, the second process, *Image Structure* (IS), searches for the individual gene spot regions within the newly found master block areas.

Whereas the IL and IS processes are focused on identifying a gene spot's local region Chapter 5 is interested in determining an accurate profile for the gene spot. With such a profile generated a more accurate determination of the gene spots intensity can be quantified. Chapter 6 explored this approach to the problem of gene spot boundary identification as implemented in *Spatial Binding* (SB). The SB process is a hybrid technique that utilizes the learnt image knowledge from the previous stages to rebuild a gene spot's characteristic topology with the help of *Vader Search* (VS).

11.2.3 Gene spot quantification

With the microarray image's gene spots accurately identified, the task becomes one of gene spot quantification. The microarray channel data for all of the gene spots found in the imagery have their florescence intensity quantified in some way. The underlying assumption is that this florescence is proportional to the amount of deoxyribonucleic acid (DNA) that bond to the underlying slide. Therefore, dye intensity is an indicator of the level of activity for the image's genes. A method is required therefore that facilitates the comparison between the control and disease genes such that a representative value is derived. By dividing the expression levels of one channel by that of the other and applying a log calculation, it is possible to track an increase and decrease in the expression of a particular gene. This quantification process is encapsulated in the previously highlighted SB process.

11.2.4   Slide and experiment normalization

Ideally, normalization is the process whereby systematic bias or trend is removed from the dataset fully. In practice it is, unfortunately, very difficult to completely remove such noise elements however. As such, microarrays normally use some form of control spots amongst the true gene spots. These control spots are known as housekeeping genes, and usually consist of genetic material from an unrelated experiment species such that they serve as quality control tests. A good example of these control spots is the Amersham ScoreCard, which contains a set of 23 artificial genes. These genes generate pre-determined signals that do not change across samples or experiments. As these "control spots" are susceptible to random noise (as per a normal gene), they must be spread throughout the array to be used effectively for normalization. Such normalization processes are encapsulated at various points (most notably the IL and IS stages) throughout the processes presented in this book.

11.3   Contributions to microarray biology domain

A microarray analysis system that is capable of processing a full size microarray image in a truly automatic fashion without the need for prior domain knowledge was proposed. This new system requires only that the input imagery be of cDNA microarray origins to begin with. The underlying system processes are able to determine an image's composition automatically in most cases, with minimal human intervention required. The technical and practical contributions in the biological domain are highlighted appropriately.

11.3.1   Technical

The technical contributions reviewed in this section aim to provide insight into the design implications of a fully autonomous microarray analysis system. First, the limitations of current microarray analysis systems are such that they are invariably manual operation heavy, suffer from a wealth of operator instigated accuracy errors and do not attempt to rectify the propagation of internal errors through the systems. The design of the CMA system took the core requirements of microarray analysis and reformulated them into dynamic framework architecture. Current systems amalgamate these analysis stages into one or two steps by consequence of either the design itself or other such domain criteria. A problem with this mentality is that too much emphasis is placed onto the individual analysis stages, therefore rendering such stages critical (single point of failure).

Limitations of current systems therefore are overcome (in most respects) with the presented CMA framework. The framework decouples traditional stages into highly optimized component parts. For example, the much utilized GenePix package allows the operator to use simplified brightness and contrast adjustments during their manual gene spot addressing task to correctly demarcate the gene spots. Such dynamic adjustment capability renders the operator a facility that should make error prone spotting decisions difficult to create. However, expecting the operator to choose an optimized view for every gene spot in the image (some 9200~12000 in test data) is obviously inappropriate. Indeed, the operator will typically choose such a setting that renders most of the gene spots balanced. Such a process is encapsulated in the ITE functionality of the framework. Although as per the human, the ITE does not optimize the view for every gene spot found (at present), the ITE process does generate far more accurate interpretations of the probable gene spot locations.

When this ITE accuracy is coupled with the PCC result, the initial ITE results are reinforced. Weak decisions generated from one component will be down weighted by for appropriate decisions in the next process. Such robustness cannot be achieved with a manual or semi-automatic process as there is too much reliance on the operator stage. The framework then, literally "falls over itself" to correct errors before they become too great.

The framework then contributes to the biological domain by offering a structured environment in which truly automatic microarray images can be analyzed. More critically perhaps, due to the framework's algorithmic decision processes a given image will always return the same analytical results (something not guaranteed with a human at the helm). Another clear benefit of the framework process is the fact that it can render the same benefits to other biological image analysis tasks.

11.3.2   Practical

A major advantage of the proposed analysis system over that of contemporary manual or semi-automatic systems is one of result repeatability. As stated throughout the book and particularly in Chapter 5, the human element of current microarray analysis systems typically introduces ambiguity into a given analysis. The task of addressing the gene spots accurately such that appropriate quantifications can be made is one example where these ambiguities can be seen for instance. The human operator of these systems will usually determine slightly differing gene spot morphology for a given gene spot over different analysis events. Such variance in the addressing stage can lead to significant analysis differences for the given gene spot at later stages. When such quantification results are then used to interpret the functionality of various genes and the interaction that takes place between them for example, such matters become ever more important.

In most cases, the proposed analysis system's reliance on internal algorithmic processes during the various stages of a typical microarray image results in a systematic decision process taking place for all objects found in the image. Such a decision process yields high repeatability for gene spot morphological identification processes (where a large part of the human requirement is focused at). As the proposed system removes the majority of the manual operator addressing task from the analysis process, the knock-on benefit of time reduction is realized. With the operator taking between 3∼4 hours to correctly address the full microarray surface on contemporary systems, any reduction in process time will be beneficial. Bearing in mind that the current system process is based on non-optimized code, the processing time for one image of 50 minutes would seem to be quite acceptable.

Indeed, remember that during these 50 minutes the operator is free to resume other research (in most cases), needing to return later simply to interpret the resultant quantifications of the underlying, microarray experiment run. Traditional analysis systems do not free up a researcher's time so dramatically.

This time reduction ability is further appreciated when it is remembered that the system can be run in a batched mode of execution. This means rather than providing one image, the operator can provide a directory structure full of microarray images that need to be processed. The system will parse the provided imagery as appropriate and render the final output similar to the biology domain standard of GenePix.

11.4 Contributions to computer science domain

The four stages as explored in Chapters 3-6 were brought together with the proposal and implementation of a framework that fully automates the processing requirements of a cDNA microarray image. Whereas techniques such as GenePix and SPOT require some form of manual demarcation task in order to identify the gene spots themselves, the proposed framework generally requires no operator intervention. This CMA framework has the potential to process not only microarray image data, but also many medical and other image sets. The framework is proving capable of processing images far more complex in surface variance than those as attempted by competing methods with the guidance of a human operator. The technical and practical contributions to the computer science domain are discussed below.

11.4.1 Technical

Clustering techniques have proved themselves to be very powerful at determining the similarity between a set of objects (with appropriate caveats). However, as modern science expands and data gathering methods improve, it is becoming more challenging to apply the techniques directly to large scale analysis problems.* The PCC process was theorized as a means by which such techniques could be directly applied to ever larger datasets. Indeed, not only does the re-scaling ability of the PCC technique facilitate the processing of these larger (twenty million–plus pixels) datasets, but the ability to render transparent a clustering processes "internal decisions" yields the possibility of creating a clustering engine that produces superior results. This superiority is most notable in the spatial domain where the PCC technique has proved to be effective in the application of microarray data.

The benefits of the proposed CMA framework to biologists working with microarray image data have been highlighted in Section 11.3. Researchers in the general field of image analysis should also benefit from the proposed framework with respect to the adaptability aspects. The CMA framework was designed specifically such that internal processes would be modular in nature. Such modularity allows the swapping out of processes as and when required depending on the task at hand. For example, the current focus on microarray image data led to the compositional algorithms examining an image for regular structures across their surface. Re-applying the framework into the 2D electrophoresis gel image domain would mean the adaptation of the compositional algorithms to highly irregular structure identification. However, due to the purposely designed modular construction, once an appropriate composition component has been crafted for gel images, this new algorithm can be executed instead of the microarray one. Such modularity gives the framework great flexibility.

11.4.2 Practical

A potential problem with current microarray analysis systems is the approach in which they solve specific tasks. Typically the systems will "lump together" functionality rather than decouple them as much as possible. A good example of this would be that the GenePix package requires the human operator to separate the gene spots from their local background by using a contrast/brightness function. Such separation therefore combines finding the gene spots with

---

*The opposite is also true, such that powerful methods (Genetic Algorithms for example) are thrown at problems simply because they can be, with the underlying hope that a better result is rendered from the GA application.

a specific contrast/brightness setting. Although this should be an ideal process to determine good separation between the two domains, due to the number of gene spots in an image, such multiple contrast/brightness settings are impractical. Typically then, the operator will choose a contrast/brightness setting that presents what looks to be the best surface view. Due to the intelligence of the human mind, the operator is then able to bring a wealth of subtle surface information to the task of determining a good border area for a given gene spot.

With the creation of a fully automatic system it could be imagined that a similar process would be easily implemented. Generally, this is the case, but much work needs to be focused at providing algorithmic insights into the subtle surface information that the mind determines subconsciously. Therefore, when creating such automatic analysis systems (be they for microarrays or other image domains), analysis processes should be decoupled from each other as much as possible. Such decoupling not only allows specific tasks to be optimized more appropriately, but also facilitates the option of providing more robust error handling capabilities. An example of this decoupling and robustness functionality can be seen particularly well with the implementation of the ITE and PCC components of the framework. Both processes are focused at a similar task (enhancing the gene spot locations), but the approach taken by the two processes are completely different. Taken individually, the results of these two components generally identify the gene spots appropriately, when used together though, such enhancements are improved. In this way, errors created by one component will be chaperoned by the results of the other, as the two approaches are unlikely to generate the same error in the same image space.

The proposed framework also opens the way for a generalized image analysis process. Although the decoupled components presented are optimized to the task of gene spot identification and tracking, the components are goal orientated. This means other image analysis problems could easily be used with modifications to these component parts. For example, imagine that the input data is from a satellite and presents battleship locations from the Atlantic Ocean. The ITE component could be extended in this case to take advantage of the multiple input sources (infrared frequencies for instance) of the imagery. This in itself would render interesting results. The PCC component would extend these image results and clarify the likely battleship locations upon the sea more clearly. However, battleship data would have to be evaluated as quickly as possible, this means the PCC process would have to be modified such that it focuses more closely on the ITE regions of interest.

A more appropriate application perhaps could be in the analysis of three-dimensional microscopic image data. Typically, these applications generate views of the object under investigation to several depths, resulting in a stack of images. These stacked images can be processed like a standard microarray surface (initially), with the only real difference to the current domain being that stack members provide more knowledge (as there are more stacks) than their microarray cousins.

11.5 Future research topics

All of the CMA framework components mentioned have performed extremely
well over the available set of microarray images. However, as is usually the
case in the research field, the techniques are not perfect in their design or
implementation at present. The following highlights certain aspects of the
components which can be investigated further in order to improve them accordingly.

11.5.1 Image transformation engine

The implemented ITE curve was derived from the results of an intensive investigation into the characteristics of a typical microarray's surface profile.
A trade off was then made such that the implemented response curve coped
well with the majority of the available image topologies. This static curve has
proved to be highly effective at reducing the bulk of a given image's artifacts.
Unfortunately, however, one of the side effects of this curve's nature is that
very weak gene spots are left "as is" on the image surface (although PCC is
able to detect weak gene spots generally, this situation could be improved).
The rationale behind this highlights that the background of the image is typically made up of weak signals (inter-dispersed with stronger artifact noise).
Importantly, the gene spot intensities of a typical microarray image exist at
the mid-to-high end of the possible 65536 values; it is conceivable that in some
areas of the image a gene spot could have an intensity range associated with
the background in another region. It is difficult to ascertain a reliable one-pass
method that will differentiate between these weak gene spots and background
noise elements (as well as any possible gene spot in fact).

One approach that could hold merit is the creation of a dynamic response
curve rather than the current static one. Dynamic is used in this context to
mean the curve responds in some way to the underlying surface. This dynamic
approach would mean the ITE process would analyze the current image surface
and generate a response curve that is focused on the gene spot frequencies
more appropriately. This new curve would have to be guided in some way (the
static curve could become a template like structure) such that large artifact
noise does not cause the algorithms to deviate significantly. In its simplest
implementation, the new response curve could be generated from a selection
of curves that highlight multiple frequencies associated with the gene spots.

A second approach could see the melding of this dynamic response curve
process with a texture analysis component. The texture of a microarray im-

age's background is substantially different from that of the gene spot foreground. This texture information should yield greater accuracy in background prediction if used correctly. When melded to the response curve generation stage, such a process would yield an ITE stage that is substantially more powerful.

11.5.2   Pyramidic contextual clustering

The PCC component has proved itself to be very powerful primarily due to its ability to render a clustering technique's internal decision process transparent. Importantly the scaling ability of the algorithm is also of significant benefit as without which an amalgamation process would have to be built. Such an amalgamation stage needs to split the image surface up into sections and then stick these sections back together after the analysis. Problems with this splitting concept include appropriate binding stages, as the interaction between the multiple sections will not have been accounted for. Rather than split the image up then, scaling the algorithms to deal with bigger datasets has many advantages. A disadvantage, however, of the *divide and conquer* (DAC) scaling approach is the added time complexity of the process. The DAC process separates the surface area up into increasingly larger distinct regions. These overlapping regions allow the PCC approach to essentially re-compute parts of the surface slightly and hence account for intensity differentials better than a standard splitting approach would.

Although the final output of the component offers substantially more knowledge of the underlying microarray image's surface, loss of gene spot information due to local background grouping becomes larger than gene spot groupings. One way to improve this feature would be the synchronization of the underlying window shape to a particular *region of interest* (ROI). Changing these dimensions to fit a problem area more closely will yield more clarity in the area. This clarification would have the knock on effect of more accurately identifying low level type features as the range of probable intensities within the area will have shifted to a local norm.

11.5.3   Image layout and image structure

The current design and implementation of the IL and IS components respectively have performed well over the available microarray image set. The individual processes have coped well with the noise present in the resultant ITE surface image, with their master and meta-block identifications being quite accurate. The multi-pass nature of these components has stood them in particularly good stead as well. However, the components assume that the gene spots have been printed in a regular rectangular grid like structure. This assumption happens to be true for the 122 images available during the development stages. This may not always be the case as there are other addressing structures available to a microarray facility. Although the functionality of the

IL and IS components can be guided somewhat by the PCC results of a given image, this guidance was designed to be superficial at best (there are after all no guarantees the PCC component has generated a perfect structure). A new approach would be to take ideas from the SB component and process these before that of the IL and IS stages.

11.5.4   Spatial binding

The greatest potential weakness of the SB process is associated with the initial gene spot tracking task. In the current implementation of the process a GHT was used to determine the initial gene spot position within the IS determined region. Although the GHT inspired position is reinforced by the gene spot ET calculation stage, a situation exists in which the GHT process can produce an erroneous position ring. This position ring causes an increase in the SB process calculation as in essence, this gene spot morphology will have to be calculated twice; once based on the GHT and once on the ET methods. These two metrics should be in general agreement and as such their resultant pixel member regions will be very similar, but a re-calculation is still required. The VS process could be further improved by casting this process back through the "time slices" of the history layers at their granularity settings to help yield further knowledge.

11.5.5   Combining microarray image channel data

At present the CMA framework is designed to only use the raw microarray image for the initial creation of the ITE imagery. Although this ITE stage has proved itself to be a powerful process, there will undoubtedly be information lost during this conversion event. cDNA microarrays at present typically consist of two embedded channels (Cy5 and Cy3 laser results respectively), although the use of four channels is also possible. Currently, microarray analysis packages typically focus on analyzing these channels independently with the results only combined at the last possible stage (the gene ratio calculation). A new approach in the analysis of these microarray images would be to harness this surface data directly. These two~four channels exhibit very high similarity and as such there are essentially two surface measures for any given pixel. This channel information should be used more explicitly, for example by generating a warp image which defines the individual pixel confidences or allegiances within a given gene spot boundary.

11.5.6   Other image sets

A key goal of the proposed CMA framework was that the internal processes should be modular. This modularity aspect would allow the easy swapping out of components as and when required depending on the task at hand. For example, the current focus on microarrays has led to the IL and IS components examining an image for regular structures across its entirety. If the framework was to be applied to 2D gel electrophoresis images this grid like structure would be highly irregular. Due to the framework's modularity however, once an appropriate component has been crafted for this task, this new component can be executed instead of the microarray one. Such modularity gives the framework great flexibility. It would be interesting to investigate on time-series images in order to explore the dynamical behaviors of gene regulatory networks [217–219].

11.5.7   Distributed communication subsystems

The development of a communication subsystem would have great benefits for the framework. For example, suppose the image structure component cannot define the appropriate structure of a PCC generated image. This issue will be directly related to the noise variance present in the image at the relative coordinates; be it negative genes, high intensity background or large artifact regions etc. Rather than take an extreme solution and abort this image re-processing the whole image via a different ITE/PCC combination, it would be preferable to reprocess just the ROI. This reprocessing would entail the IS requesting that the ITE component regenerate the ROI with a different emphasis. For example, if the ROI was high in negative genes the ITE could simply invert it; alternatively the ITE could harness information from the PCC generated image and reprocess those objects within the ROI.

12

Appendices

12.1 Appendix A: Microarray variants

As can be imagined, there are several competing technologies that facilitate the simultaneous examination of thousands of genes. Appendix A highlights some of the main contenders in this realm, along with relevant material. The digital generation stage of the microarray process is usually completed with a specially built scanner device; the key differences between the two most widely used scanner systems are highlighted appropriately.

12.1.1 Building the chips

Although there are several distinct steps in the microarray chip building scenario, the first step typically begins with choosing an appropriate substrate, which has predictable and reproducible characteristics. To avoid a multitude of process difficulties with regards to chemicals and toxicology, however, pre-treated slides with well proven protocol methods are normally purchased from third parties to reduce variability problems. With this base material acquired, the next step is to spot the genetic material of the targets onto the surface of the pre-printed probes on the slide. Deoxyribonucleic acid (DNA) chip or "Chip fabrication" for a microarray can take one of two different approaches. Either being printed directly onto the base material (as stated in the scenario) via a robotic system or essentially grown via photolithographic techniques. In the case of a glass substrate base material (the most common choice), these systems can also be grouped according to the attachment chemistry used to bond the probe to the slide.

12.1.1.1 Robot based

Robot or complementary deoxyribonucleic acid (cDNA) based glass slide [185] systems are constructed in such a way that they contain three distinct areas. The first consists of the robotic arm and controller. This arm unit is normally a static matrix structure which is attached to the outer shell of the microarray device. In an ideal situation, during an experiment run this arm is the only component that will move within the microarray. However, this is not always the case as the microarray itself is rarely bolted to a stable carrier.

The second area is commonly called the hotel. This area contains the well plates (384 wells are typical) where the mixed samples sit in preparation for the spotting process. There is also a "head cleaner" in this area which allows the matrix structure to be washed when required as blockages and the like materialize from time to time.

The third area being the largest consists of the "glass slide" bays which the targets are deposited onto. The spotting process in this situation would proceed along the lines of 1) move matrix to hotel and acquire sample, 2) move matrix to slide bay and deposit sample, 3) move matrix to wash unit and clean heads, and 4) repeat cycle from position 1. Depending on the pin construction type and the printing technology used, these microarray systems can deposit between 200 and 30,000 genes per chip area of $1cm^2$.

Robot based nylon-membrane systems are typically more sensitive than their glass slide counterparts and are still in use in some laboratories. The advantage of such systems would appear to be their Southern Blot affinity and as such molecular biologist acceptance. These systems use radioactive based labeling (due to health issues, non-radioactive methods exist) as this greatly increases* the sensitivity of probe detection over fluorescent based systems. However, due to the porous nature of the membrane material these systems are used on organisms with small genomes or on subsets of larger organisms. Also, due to the nature of the labeling process, the omni-directional emissions from a gene spot can co-mingle with neighboring spots in greater propensity. This emission based information is captured by a phosphor imaging system, which is then converted to the digital realm by a traditional scanner device.

Fluid Arrays as described by Biran et al. [29, 75] represent a fundamental shift in microarray processing methodology. These Fluid array systems use etched optical fiber bundles which are individually filled with oligonucleotide-functionalised micro-spheres16. These bundles consist of thousands of individually addressable fibers which enable parallel detection. Specific hybridization is detected by fluorescence only at probe positions complementary to their targets. The probe hybridization result is passed by the fiber cable to traditional detection equipment. Because each fibre position must contain one probe, replicates for each position must be created. Also, due to the random distribution of the array surface there needs to be a very accurate labeling system in place.

12.1.1.2 Photolithograph based

In contrast to the robot based systems, Fodor [78, 79, 163] developed photolithographic or oligonucleotide systems which took advantage of methods applied in the semiconductor industry. Rather than deposit a pre-mixed solution of probes onto a pre-printed glass slide, these systems build the spe-

---

*Fluorescence based linear range typically three log spread rather than radioactive based four to five log spread.

cially selected probes onto the glass slide on a base by base basis. Here, the sugar of the hydroxyl group of the current nucleotide is covalently bonded with the phosphate group of the next nucleotide in the sequence. Every nucleotide added has a protective group on its 5' position to prevent multiple base attachments in a round of synthesis. Once a round has completed, this protection group is removed via a specially designed mask. This mask allows UV light to pass only to some areas of the array such that subsequent bases can be added in the next round. In the case of the Affymetrix [2] implementation of this technology, Affymetrix defined (as compared to their robot based counterparts) a rigid set of protocols. These protocols essentially construct a biotin-labeled complementary RNA 16 built up base by base using conventional phosphoramidite based chemistry or with pre-formed oligonucleotides. As these protocols are highly defined the comparison of multiple arrays should be a relatively straightforward task.

Photolithographic systems are perhaps the most powerful as the nucleotide sequences can be built from any appropriate repository (the human genome project for example). However, herein lies a weakness with this construction method, such that the sequences are limited to approximately twenty five base pairs (Newer devices can work with seventy base pairs). The masks are also very expensive to construct, with all iterations of the synthesis requiring a new mask. Of course, large numbers of identical arrays can easily be produced once the mask set has been constructed. In contrast to using a mask for the protection of certain areas, mask-less systems rely on a set of micro-mirrors which reflect the light to the appropriate part of the surface.

12.1.1.3 Attachment chemistry

For cDNA based probes the attachment chemistry can either be covalent or non-covalent. In covalent based attachment, an aliphatic amine (NH2) group is added to the probes such that this group can then covalently bond to chemical markers on the pre-prepared glass slide. With non-covalent attachment, a probe's phosphate backbone is electro-statically attracted to the aliphatic amine group on the glass slide. The non-covalent approach is unstable for oligonucleotide based probes as by definition these probes have small sequence lengths and thus a small number of backbone attachment points.

12.1.2 Digital generation

12.1.2.1 Charge coupled devices

A charge coupled device (CCD) is a collection of tiny light-sensitive diodes, which convert photons (light) into electrons (electrical charge). These diodes are called photosites and each one is sensitive to light, the brighter the light that hits a photosite, the greater the electrical charge that will accumulate at that site. The CCD along with mirrors, lenses and filters make up what is called the scan head, which moves across the surface of the slide. This head arrangement is also secured to a stabilizer bar to ensure that there is no wobble or deviation in a given pass. Essentially a lamp sits over the CCD device with the image of the slide reflected by one angled mirror which in turn reflects into another mirror. The mirrors are slightly curved to focus the image they reflect onto a smaller surface. The last mirror reflects the image onto a lens, with this lens re-focusing the image onto the CCD array. As charge coupled devices do not capture any color information, filters can be used. The lens splits the received image into three smaller versions of the original. Each version passes through a color filter (red, green or blue) onto a discrete section of the CCD array. The scanner combines the data from the three parts of the CCD array into a single full-color image. Aspects of so-called CCD generated noise should be accounted for (partially at least) during the scanner setup stage within individual laboratories. However, such local calibration processes are usually focused on the chemical usage and other biological protocols required for the underlying biological mechanisms themselves. Perhaps an appropriate calibration stage for such CCD anomalies would be that of flat fielding. Due to imperfections in the manufacturing process the sensitivity of the pixels will vary slightly (usually a few percent) across the CCD grid. This effect is essentially random, and is not specifically a function of the pixels grid position, for example. The CCD pixel's relative sensitivities can be calibrated by imaging an evenly illuminated source, an ideal blank slide for instance and examining the variation in pixel values recorded. Once this calibration is known, microarray images can be corrected to the values they would have had if all the pixels had been uniformly sensitive.

12.1.2.2 Con-focal

The most sensitive systems (and the most common) use a con-focal beam of light that is shrouded such that the returned signal is limited to that of the slide surface itself. Any signal returned from other locations around the slide will be slightly out of focus and therefore will be weaker in a photonic sense. Since the dyes used to fluorescently tag the nucleotides react to different light frequencies, the lasers are tuned to specific wavelengths. As previously stated, the dyes commonly used in two-color microarrays are known as Cyanine 5 (Cy5) and Cyanine 3 (Cy3). These have wavelengths of 635nm and 532nm respectively. These reflected photons are then detected by a photomultiplier tube (PMT) with an analogue to digital (A/D) converter turning these electrons into their digital equivalent. This digitization averages the input signal both spatially and temporally and thus produces as output an average measure of florescence (and therefore transcription abundance) for a given pixel location. The setting of the PMT and on some scanners the lasers themselves are critical to good image production. Generally, higher laser power settings generate more signal-bleaching and source noise, while higher PMT settings generate more electrons per photon which in turn creates more detector noise and signal. These settings are critical as during the conversion step, the upper bound for an acceptable signal is typically 65535 or 216-lb-bits. This has the effect of saturating pixels and losing accurate gene information. The scanner devices must therefore be calibrated such that the minimum and maximum detected signals from the image are represented by 0-65535 respectively. This scanning process operates in a similar manner to that of the CCD method, with the laser then moving across the surface of the slide acquiring these intensities and converting them to digital equivalents as it goes.

In the case of membrane systems, the phosphor image surface is substantially larger than the equivalent glass slide or photolithographic surface. Therefore con-focal based scanning of these systems typically trade resolution quality for scan time.

12.2 Appendix B: Basic transformations

The hybrid frequency response transformation as proposed in Chapter 3 is based on three fairly simple mathematical mapping functions. Although these functions are not new per se, their direct usage in microarray image enhancement is minimal. Chapter 3 showed that such transforms can be a basis for fairly powerful mapping processes.

12.2.1 Linear transform generation

The simplest form of remapping that can be applied to a given image surface is that which renders a linear relationship. Here, the input range of 0-65535 is linearly rescaled to the range of 0-255. This linear conversion will provide the simplest and quickest representation of a microarray's intensity range with large amounts of detail being discarded in the process. Mathematically, this function is presented in Equation (12.1). The linear response curve (or the mapping path) is presented in Figure 12.1 with the appropriate raw and remapped surface areas presented in images in Figure 12.2 respectively.

$$L(x) = \frac{x}{2^8} \qquad (12.1)$$

where x is the 16-bit intensity value that is converted into 8-bits.

The left image in Figure 12.2 gives the impression that there is very little background noise present, but as shall be seen in the coming discussion this is false. From the right image in Figure 12.2 it can be seen that the linear transformation has indeed retained the general intensity information of the surface area. However, a significant number of gene spots are indistinguishable from their background. Although this linear scaling results in the least amount

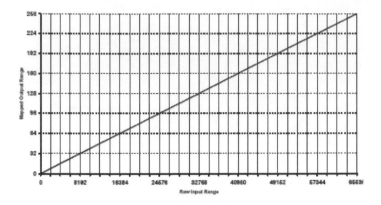

FIGURE 12.1
Linear response curve mapping

FIGURE 12.2
Transformation results. Left: original input image; Right: mapped output image

of distortion to the microarray image, the gene spot intensity ranges are not distributed evenly throughout the 16-bit range. The mapping process should put greater emphasis on the mid to high level intensity values, as this is where the gene spots reside in abundance. This example shows that the reduction process of a one-to-one remapping is viable, but the benefits of the scheme are negligible in this linear case. Therefore, a scheme must be used that retains this reduction aspect, but at the same time, enhances one or other element of the surface.

12.2.2 Square root transform generation

The square root transform represents a good compromise choice of mapping technique that retains the reduction and enhancement aspects as mentioned above. Here the gene spot and background intensities are captured in equal amounts with the results more detailed in low to mid range intensities. Intensity information could still be lost in the process, but these lost intensities reside in regions more specific to background. This square root function is presented in Equation (12.2). The square root response curve is presented in Figure 12.3 with the appropriate raw and remapped surface areas presented in images in Figure 12.4 respectively.

$$S(x) = \sqrt{x} \qquad (12.2)$$

where x is the 16-bit intensity value that is converted into 8-bits.

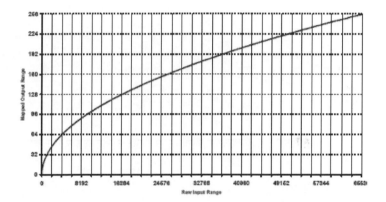

FIGURE 12.3

Square root response curve mapping

FIGURE 12.4

Transformation results. Left: original input image; Right: mapped output image

From the plot of Figure 12.4, it can be seen that the mapping emphasis has been modified in the lower or left hand side of the plot. These lower range intensity values are typically representative of background regions of an image rather than the mid-to-high range values typically associated with gene spots. The remapped image (right image of Figure 12.4) shows this clearly in comparison to that of the right image of Figure 3.2. In most cases, the gene spots have been captured along with their local background regions. The downside of this mapping is that low level gene spots can be classed as background and conversely, high level artifacts can be classed as gene spots. Essentially, the square root function retains more of the surface variation of a gene spot, along with prominent background artifacts.

12.2.3   Inverse transform generation

The square root transform represents a good compromise mapping between a surface's foreground and background intensities. The inverse transform performs rather well when the region of interest is the background. Here, greater emphasis has been placed on low intensity ranges with the results highlighting the background region of the image rather than the gene spots. Equation (12.3) presents the transform function mathematically.

$$I(x) = 1 - \left(\frac{1}{\frac{x}{2^8} + 1} \right) 2^8 \tag{12.3}$$

where x is the 16-bit intensity value that is converted into 8-bits.

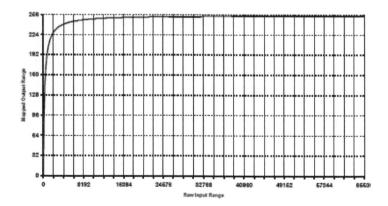

FIGURE 12.5
Inverse response curve mapping

The square root plot gave emphasis to the low to mid range intensities, as can be seen from the inverse transform plot in Figure 12.5. The inverse transform heightens the low to mid range intensities much more. As can be seen from the right image of Figure 12.6, the background intensity values are more prominent than those of the gene spots.

FIGURE 12.6

Transformation results. Left: original input image; Right: mapped output image

12.2.4 Movement transform generation

As emphasized in the right image of Figure 12.8, the background region of a microarray image can fluctuate quite a bit across the whole surface area. It would seem however that this fluctuation is quite static in nature with relatively consistent intensities throughout. If however this movement is investigated further, it becomes quite obvious that this static nature is false, as there are large intensity fluctuations present. Equation (12.4) presents the transform function mathematically.

$$M(x) = \text{significant} \left(\frac{x}{2^8} \right) \tag{12.4}$$

where x is the 16-bit intensity value that is converted into the significant 8-bit value.

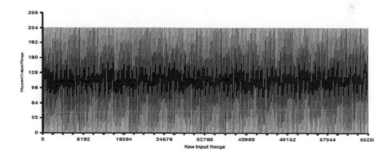

FIGURE 12.7

Movement response curve mapping

This example shows clearly in the right image of Figure 12.8 that the movement in the background intensities are not consistent as suggested in Figures 12.7-12.8. There are in fact several regions of the image that have strong biases to either the red or green image channels as can be seen by the intensity

FIGURE 12.8
Transformation results. Left: original input image; Right: mapped output image

distributions. It is important to note that it is only the top part of the image that has a consistent area. Ultimately, the smooth transition seen in this background area should be seen across the whole surface. This variability means it will be very difficult to generate a mapping that is perfect. However, the mapping does not need to be perfect as long as a significant proportion of the gene spots are not lost in the conversion process.

12.3   Appendix C: Clustering

Chapter 4 proposed a novel clustering process that uses the spatial information imbedded into the data to acquire more knowledge. This new process not only scales traditional clustering methods to ever increasing datasets sizes, but also gives insight into a clustering process's internal decision process. A brief review of clustering as it pertains to research in Chapter 4 is presented in this appendix.

Given a gene expression matrix, clustering methods are often used as the first stage in data analysis. The goal of clustering here is to group sets of genes together that share the similar gene expression pattern across a dataset.

There are three essential steps in a typical pattern clustering activity:

- *Pattern representation* refers to things such as the number of available patterns, the number of features available to the clustering algorithm, and for some clustering methods the desired number of clusters. Suppose we have a gene expression dataset with 1000 genes, whose expression levels have been measured for 10 time points or 10 different samples. In this case, we would have 1000 available patterns and 10 features available for each pattern. One clustering aim would therefore be to see how we could group these 1000 patterns using these 10 features into a number of *natural* clusters.

- *Pattern proximity* is usually measured by a distance (dissimilarity) metric or a similarity function defined on pairs of patterns. The Euclidean distance is probably the most commonly used dissimilarity measure in the context of gene expression profiling. On the other hand, a similarity measure such as such as Pearson's correlation or Spearman's Rank Correlation is also used. Other distance metrics include Mahalanobis Distance, Chebyshev distance and Canberra metric.

- *Pattern grouping* or clustering methods may be grouped into the following two categories: hierarchical and non-hierarchical clustering [110]. A hierarchical clustering procedure involves the construction of a hierarchy or tree-like structure, which is basically a nested sequence of partitions, while non-hierarchical or partitioned procedures end up with a particular number of clusters at a single step. Commonly used non-hierarchical clustering algorithms include the k-means algorithm, graph-theoretic approaches via the use of minimum spanning trees, evolutionary clustering algorithms, simulated annealing based methods as well as competitive neural networks such as Self-Organising Maps.

It is worth pointing out that one of the most powerful methods available when analyzing microarray data (or most data for that matter) is that known as clustering [109, 110, 133]. For a detailed review of clustering algorithms see Xu [225]. A loose definition of clustering could be the process of organizing objects into groups whose members are similar in some way to each other. A cluster is therefore a collection of objects which have some similarity between them and are dissimilar to the objects belonging to other clusters. The clustering process for microarray expression data works on the principle of "guilt by association" [92, 119, 196], which is to say, the unsupervised grouping of objects (genes in this case) into clusters based on a similarity in their biological function. Occasionally, this "guilt by association" concept leads to the identification of previously unknown gene functions. For example, if two genes are consistently grouped together by a clustering algorithm (they have similar functionality) and if the function for this cluster has been previously identified, then by association, the other genes in the same cluster may have the same function. Downstream analysis can then be focused onto these "similar" genes specifically. In Jain [109], the authors defined five fundamental stages of a clustering process including pattern representation, pattern proximity, the actual clustering, data abstraction and an assessment of the clustering result.

Fundamentally, Pattern Representation (including feature extraction and selection) refers to such items as number of classes and patterns available to identify the objects to be clustered, as well as the number, type and scale of these objects. Feature Extraction deals with one or more transformations of the input objects to produce new salient features, as opposed to Feature Selection which is quite obvious. For example, a microarray image could be stored using a single bit per pixel with the resultant output being rendered in black and white. Clearly, the original microarray image would have many levels of intensity available, but this information would have been lost with the above storage schema. Obviously this is an extreme example, but in general the degree to which data loss affects the output could lead to a bad representation and consequently unexpected results. Generally, a good representation is an n×d pattern matrix where there are n objects with d attributes associated with them. In the case of microarray expression data, n could be the number of genes and d could be the number of samples for each gene. Alternatively for microarray image data, n could be the number of foreground objects with d being their largest intensity measure. The attributes could be quantitative (continuous, discrete or interval values), qualitative (nominal or unordered), or ordinal, based on the objects at hand.

Once a good representation of the data is present, some method is required that allows the characteristics or attributes between the various data objects to be compared. This method is technically known as Pattern Proximity and is usually used to mean measuring the distance between two objects in the set. Typically this distance measure is used to measure the dissimilarity that exists between the objects, whereas similarity measures do the opposite [109, 110, 225].

The actual process of clustering or grouping the data objects themselves comes in many different forms over a variety of flavors. Figure 12.9 presents a possible axonometric representation of these methodologies with emphasis on the techniques used in the Pyramidic Contextual Clustering (PCC) component of the Copasetic Microarray Analysis (CMA) framework.

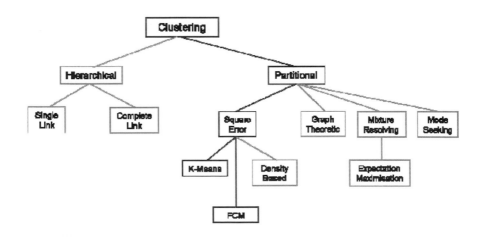

FIGURE 12.9

A sample axonomy of clustering approaches

This diagram does not try to represent all possible types of clustering approaches that are available, but simply tries to impart an appreciation of the families of familiar techniques. Note also that emphasis has been placed on the methods as used by the current implementation of the PCC component.

As shown, clustering algorithms can be broadly classified into methods that produce hierarchical or partitioned results. Hierarchical methods include single-link [192], complete-link [122] or minimum-variance [147, 213] methods. Hierarchical methods produce a nested series of partitions (based on some criteria) for merging and splitting the clusters based on a similarity or dissimilarity score. Partitioned methods have advantages in applications with large datasets for which construction of the dendrogram (hierarchical process) is computationally expensive. Such partitioned methods identify a partition that optimizes (typically in a local sense) a given clustering criterion. The most frequently used criterion function in partition techniques is that based on determining the squared error between objects. In this context the k-means clustering algorithm [139] and its variants [10] are the most common in use for hard partitioning. Soft partitioning, which was proposed by Zaden [229] and later implemented by Ruspini [176], is represented by the Fuzzy c-means partitioning techniques.

Data abstraction is the process of extracting a simple representation of the dataset. Typically this abstraction is presented as representative patterns such as the clusters centroid [59, 109, 110]. This pattern representation works well when the resultant clusters are compact or isotropic. However, when the clusters are elongated or non-isotropic, the centroid scheme fails to represent the clusters adequately. In this case, the use of a collection of boundary points to capture the shape of the cluster is more appropriate.

All clustering algorithms will, when presented with data, produce clusters, regardless whether the data contains clusters or not. Therefore some form of assessment is required that can categorize a cluster as "good" or "bad" for the purposes of the task at hand. There is little in the way of gold standards for this assessment process as the usual criterion is a one of optimality (based upon some subjective quality). Therefore validity assessments are objective [63] and are performed to determine whether a cluster result is meaningful. Clustering results are therefore considered valid if the result could not reasonably have occurred by chance or as an artifact of the algorithm itself.

Two standard clustering algorithms, k-means and fuzzy c-means, are used throughout the microarray domain. These two clustering methods use variations of a squared error criterion function. The generic squared error SE formula for a cluster set C of an object set X (which contains K clusters) is shown in Equation (12.5) as follows:

$$\mathrm{SE}(X, C) = \sum_{j=1}^{K} \sum_{i=1}^{n_j} \|x_i^{(j)} - c_j\|^2 \tag{12.5}$$

where $x_i^{(j)}$ is the ith object belonging to the jth cluster and c_j is the centroid of the jth cluster.

The k-means and fuzzy c-means algorithms are described as follows. To facilitate the readers, we will also briefly discuss the traditional hierarchical clustering algorithm.

12.3.1 K-means algorithm

Essentially the k-means algorithm is initialized with a random partition. The partitions underlying patterns are constantly re-assigned to different clusters (based on some similarity measure) until a convergence criterion is met. Typically, this convergence criterion is crafted such that the process is halted when no further pattern re-assignments are made or the squared error SE, ceases to decrease significantly.

A major issue with the SE algorithm is its sensitivity to the initial partition's selection, with a poor choice resulting in the convergence function focusing on a local minimum value. The algorithm also has an issue where a cluster can be generated such that there are no objects associated with it (the initial cluster centers are distant from an actual object for example). Strictly speaking, there are many versions of the k-means algorithm, not necessarily with that name. Similar algorithms were introduced in 1965 by Forgy [80], Ball [19], Jancey [112] and in 1969 by McQueen [139]. These algorithms were developed for statistical clustering applications, the selection of a finite collection of templates that well represent a large collection of data in the MSE sense. McQueen used an incremental incorporation of successive samples of a training set to design the codes, each vector being first mapped into a minimum distortion reproduction level representing clusters and then the level for that cluster being replaced by adjusted centroids. The pseudo-code algorithm presented in Algorithm 12.1 is an implementation of the Forgy version of the k-means technique which has several advantages over the others. Indeed there is experimental evidence by Larsen and Steinbach [125, 198] that compared the various techniques and showed that the Forgy technique frequently yields better results.

```
 1  Input
 2  X: Objects to Cluster
 3  K: Number of Clusters
 4
 5  Output
 6  C: Cluster Centroids
 7  M: Cluster Membership
 8
 9  Algorithm
10  Function k-means(X,K,C,M)
11      While not convergent
12          Choose K cluster centers to coincide with K randomly
                chosen objects
13          Assign each pattern to the closest cluster center C
14          Reassign C using current cluster M
15      End
```

```
16 End Function (C,M)
```

Algorithm 12.1: *K*-Means Algorithm Pseudo-Code

12.3.2   Fuzzy *c*-means algorithm

Whereas traditional clustering techniques generate partitions that house the underlying patterns uniquely (a pattern can only belong to one cluster), the fuzzy *c*-means (FCM) approach extends this notion of partitions by associating an object to a cluster with a membership function [229]. The membership function values are assigned in the range of [0, 1] for each cluster. Larger membership values indicate higher confidence in the assignment of an object to a cluster. Therefore, the FCM resultant clusters represent a fuzzy set of the objects rather than a partition per se. Although the FCM algorithm is better than the *k*-means algorithm for avoiding the trap of local minima, it can still converge to local minima of the SE criterion. Several variants of the FCM technique have modified the membership functions such that detecting circular and elliptical boundaries [53] is possible. A pseudo-code implementation is presented in Algorithm 12.2 as follows:

```
 1 Input
 2 X: Objects to Cluster
 3 K: Number of Clusters
 4
 5 Output
 6 C: Cluster Centroids
 7 M: Cluster Membership
 8
 9 Algorithm
10 Function Fuzzy c-means(X,K,C,M)
11     While C change not significant
12         Choose initial fuzzy partition of X objects into K
                 clusters
13 by selecting XxK membership matrix C
14         Find criterion function value in C (se for example)
15         Reassign C using current cluster C
16     End
17 End Function (C,M)
```

Algorithm 12.2: Fuzzy *c*-Means Algorithm Pseudo-Code

12.3.3   Hierarchical clustering

The concept of the hierarchical representation of a dataset was primarily developed in biology. Therefore it is no surprise that many biologists have found it easier to use hierarchical methods to cluster gene expression data. Basically a hierarchical algorithm will produce a hierarchical tree or dendrogram representing a nested set of partitions. Sectioning the tree at a particular level leads to a partition with a number of disjoint clusters. A numerical value, essentially a measure of the distance between two merged clusters, is associated with each position up the tree where branches join.

There are two main classes of algorithm for producing a hierarchical tree. An agglomerative algorithm begins with placing each of the n available gene expression values across a set of arrays (patterns, also known as the gene expression vector) into an individual cluster. The algorithm proceeds to merge the two most similar groups to form a new cluster, thereby reducing the number of clusters by one. The algorithm continues until all the data fall within a single cluster. A divisive algorithm works the other way around: it operates by successively dividing groups beginning with a single group containing all the patterns, and continuing until there are n groups, each of a single individual. Generally speaking, divisive algorithms are computationally less efficient.

A typical hierarchical agglomerative clustering algorithm is outlined below:

1) Place each pattern (gene expression vector) in a separate cluster.

2) Compute the proximity matrix of all the inter-pattern distances for all distinct pairs of patterns.

3) Find the most similar pair of clusters using the matrix. Merge these two clusters into one, decrement number of clusters by one and update the proximity matrix to reflect this merge operation.

4) If all patterns are in one cluster, stop. Otherwise, go to the above Step 2.

The output of such algorithm is a nested hierarchy of trees that can be cut at a desired dissimilarity level forming a partition. Hierarchical agglomerative clustering algorithms differ primarily in the way they measure the distance or similarity of two clusters where a cluster may consist of only a single object at a time.

A challenging issue with hierarchical clustering is how to decide the optimal partition from the hierarchy, i.e. what is the best number of groups? One approach is to select a partition that best fits the data in some sense, and there are many methods that have been suggested in the literature. It has also been found that the single-link algorithm tends to exhibit the so-called chaining effect: it has a tendency to cluster together at a relatively low level objects linked by chains of intermediates. As such, the method is appropriate if one is looking for "optimally" connected clusters rather than for homogeneous spherical clusters. The complete-link algorithm, on the other hand, tends to produce clusters that tightly bound or compact, and has been found to produce more useful hierarchies in many applications than the single-link algorithm. The group-average algorithm is also widely-used. Detailed discussion and practical examples of how these algorithms work can be found in [110].

12.3.4 Distances

As stated in Section 4.2, object or pattern proximity is a fundamental requirement of the clustering paradigm. Due to the large variety of possible feature
types and scales, the distance measure must be chosen appropriately. Such
measures are calculated by measuring the proximity (or similarity, dissimilarity) between two objects within a given feature space. The Minkowski family
of distance metrics represent commonly used metrics and are of the form as
shown in Equation (12.6):

$$\text{MD}_{ij} = \left(\sum_{l=1}^{d} |X_{il} - X_{jl}|^n \right)^{1/n} \tag{12.6}$$

Typically, features with large values and variances tend to dominate other
problem space features.

The Euclidean distance has special appeal as it is commonly used to evaluate object proximity in both two and three-dimensional space (this is not
to say Euclidean distance is limited to two or three dimensional space). The
measure itself is equivalent to a straight-line drawn between the pair of objects. Given the intensity values observed in the microarray image at k points
between two pixel members (x_i, x_j), the formula for calculating the Euclidean
distance, ED, is presented in Equation (12.7):

$$\text{ED}_{ij} = \left(\sum_{l=1}^{d} |X_{il} - X_{jl}|^2 \right)^{1/2} \tag{12.7}$$

ED is a special case of the Minkowski distance at $n = 2$, and is invariant to
translation and rotation. This metric tends to form hyperspherical clusters.

City-block distance is a special case of the Minkowski metric at $n = 1$.
The City-block form tends to generate hyperrectangular clusters. Given the
intensity values observed in the microarray image at k points between the two
pixel members (x_i, x_j), the formula for calculating the City-block distance, CB
is presented in Equation (12.8):

$$\text{CB}_{ij} = \sum_{l=1}^{d} |X_{il} - X_{jl}| \tag{12.8}$$

12.4 Appendix D: A glance on mining gene expression data

DNA microarray technology has enabled biologists to study all the genes
within an entire organism to obtain a global view of gene interaction and

regulation. This technology has great potential in obtaining a deep understanding of biological processes. However, to realize such potential requires the use of advanced computational methods. In the last chapter, experimental protocols and technical challenges associated with the collection of gene expression data were presented, the nature of the data was highlighted, and the data standards for storing expression data were discussed. In this chapter, we shall introduce some of the most common data mining methods that are being applied to the analysis of microarray data and discuss the likely future directions.

Data mining has been defined as the process of discovering knowledge or patterns hidden in (often large) datasets. The data mining process is often semi-automatic and interactive, but there is a great desire for it to be automatic. The knowledge or patterns discovered should be meaningful and interesting from a particular point of view. For example, loyalty patterns of previous customers should allow a company to detect among the existing customers those likely to defect to its competitors. In the context of DNA microarray data, the patterns detected should inform us of the likely functions of genes, how genes may be regulated, and how they interact with each other in health and disease process. A comprehensive treatment of various issues involved in building data mining systems can be found in Hand et al. [99].

12.4.1   Data analysis

The job of a data analyst typically involves problem formulation, advice on data collection (though it is not uncommon for the analyst to be asked to analyze data which have already been collected, especially in the context of data mining), effective data analysis, and interpretation and reporting of the findings. Data analysis is about the extraction of useful information from data and is often performed by an iterative process in which "exploratory analysis" and "confirmatory analysis" are the two principal components.

Exploratory data analysis, or data exploration, resembles the job of a detective: understanding the evidence collected, looking for clues, applying relevant background knowledge and pursuing and checking the possibilities that clues suggest. This initial phase of data examination typically involves and number of steps, namely,

- *Data quality checking*, processes for determining the levels and types of noise (random variation) in the system, processes for dealing with extreme outlying values and missing values.

- *Data modification*, processes for transforming data values (such as the log2 transformation and processes for error correction.

- *Data summary*, defined methods for producing tables and visualization of summary statistics for the data.

- *Data dimensionality reduction*, methods for reducing the high dimensionality of the data to allow patterns to be identified or other data mining tools to be used.

- *Feature selection and extraction methods* that allow defining features of the data to be identified and significance values applied to such features.

- *Clustering methods*, such as hierarchical clustering, which identify potential structures in the data the share certain properties.

Data exploration is not only useful for data understanding, but also helpful in generating possibly interesting hypotheses for further investigation, a more "confirmatory" procedure for analyzing data. Such procedures often assume a potential model structure for the data, and may involve estimating the model parameters and testing hypotheses about the model.

12.4.2 New challenges and opportunities

Over the last two decades, we have witnessed two phenomena which have affected the work of modern data analysts more than any others. First, the size of machine-readable datasets has increased and the problem of "data explosion" has become apparent. Many analysts no longer have the luxury of focusing on problems with a manageable number of variables and cases (say a dozen variables and a few thousand cases); and problems involving thousands of variables and millions of cases are quite common, as is evidenced in biology throughout this book.

Second, data analysts, armed with traditional statistical packages for data exploration, model building, and hypothesis testing, now require more advanced analysis capabilities. Computational methods for extracting information from large quantities of data, or "data mining" methods, are maturing, e.g. artificial neural networks, Bayesian networks, decision trees, genetic algorithms, statistical pattern recognition, support vector machines, and visualization. This combined with improvements in computing such as increased computer processing power and larger data storage device capacity both at cheaper costs, coupled with networks providing more bandwidth and higher reliability, and On-Line Analytic Processing (OLAP) all allow vastly improved analysis capabilities.

These improvements are timely as functional genomics data provides us with a problem where traditional mathematical or computational methods are not really suited for modeling gene expression data: such data are noisy and high-dimensional with up to thousands of variables (genes) but a very limited number of observations (experiments). Some of the most commonly used methods for gene expression data exploration are hierarchical clustering, k-means (as discussed in Appendix D), and self-organizing maps (SOM). Also, support vector machines (SVM) have become popular for classifying expression data.

12.4.3   Data mining methods for gene expression analysis

To date two main categories of data mining methods have been used to analyze gene expression data: clustering and classification. Clustering is about the organization of a collection of unlabeled patterns (data vectors) into clusters based on similarity, so that patterns within the same cluster are more similar to each other than they are to a pattern belonging to a different cluster. It is important for exploratory data analysis since it will often reveal interesting structures of the data, thereby allowing formulation of useful hypotheses to test. In the context of gene expression data analysis, clustering methods have been used to find clusters of co-expressed/co-regulated genes which can be used to distinguish between diseases that a standard pathology might find it difficult to tell apart.

As early DNA microarray experiments have shown that genes of similar function yield similar expression patterns, work on the use of classification techniques to assign genes to functions of a known class has increased. Instead of learning functional classifications of genes in an unsupervised fashion like clustering, classification techniques start with a collection of labeled (pre-classified) expression patterns; the goal being to learn a classification model that will be able to classify a new expression pattern. Classification has also been extensively used to distinguish (classify) different samples, for example, breast cancer samples into distinct groups based on their gene expression patterns. For more details please refer to [133].

12.5   Appendix E: Autocorrelation and GHT

Chapters 5 and 6 call upon traditional as well as new methods to accomplish the successful identification and separation of gene spots in a given microarray image surface. Appendix E mentions the key features of these techniques and gives links to further reading.

12.5.1   Autocorrelation

Source material copyrighted to http://en.wikipedia.org/wiki/Autocorrelation.

Autocorrelation is a mathematical technique used in signal processing in order to analyze functions or a series of values (time domain signals for example). Specifically autocorrelation is the cross-correlation of a signal with itself. The autocorrelation technique is useful when searching for a repeating pattern within a given signal. For example, when determining the presence of a periodic signal swamped with broadcast noise or to identify the fundamental frequency of a signal. In signal processing, autocorrelation can give information about repeating events like musical beats or pulsar frequencies, even though it cannot tell the position in time of the beats themselves.

Figure 12.10 shows the result of applying autocorrelation to the first 4 seconds of a MIDI (musical instrument digital interface) encoded sound file (The Blue Danube) in the plot of the left image and the resultant autocorrelation sequence of this signal in the plot of the right image.

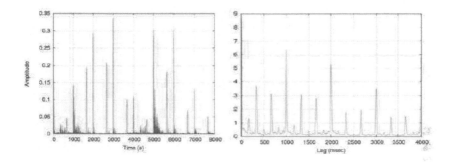

FIGURE 12.10

Example autocorrelation process. Left: original signal; Right: autocorrelation of signal (first 4 seconds)

12.5.2 Generalized "circular" Hough transform

Source material copyrighted to http://www.netnam.vn/unescocourse/computervision.

The generalized Hough transform is a method for locating instances of a known pattern (a circular region in this context) within an image. The search pattern is parameterized as a set of vectors from feature points in the pattern to a fixed reference point. Typically, these feature points are edge features and the reference point is normally the centroid of the search pattern. To locate the pattern in an image the set of feature points in the image are considered. An accumulator array keeps track of the frequency with which each possible reference point location is encountered. When all of an image's features have been processed the accumulator array will contain high values (or peaks) at the locations where image features coincided with pattern features. The Hough transform can detect multiple similar objects in one pass. It can also be enhanced to locate instances of a pattern at different orientations and scales. In this case, a four dimensional accumulator array is required with the appropriate computational complexity increasing by two orders of magnitude.

When the object of investigation is to search for circles of known radius R, the identity equation (12.9) can be used:

$$(x - a)^2 + (y - b)^2 = R^2 \tag{12.9}$$

where (a, b) is the center of the circle.

Essentially a circle is incremented in the (a, b) accumulator array center at radius R for each edge pixel to be processed. Eventually, the highest values in this accumulator fill the (a, b) array, indicating coincident edges in (a, b) space corresponding to a number of pixels on the edge of the same circle in space. Algorithm 12.3 presents the pseudo-code for such a generalized Hough transform.

```
1  Input
2  I: Raw image surface
3  Ref: reference diameter
4
5  Output
6  A: accumulator
7  SC: Shape centroids
8
9  Algorithm
10 Function GHT(I,Ref,A,SC)
11     Find all desired feature points in space
12         For each feature point
13             For each pixel I on the target boundary
14                 Get the relative position of reference point
                        from i
15                 Add this offset to the position of i
16                 Increment that position in accumulator A
17             End
18         End
19         Find local maximum in accumulator A
20     End
21 End Function (A,SC)
```

Algorithm 12.3: Generalized Hough Transform Algorithm Pseudo-Code

The following points should be remembered when using such techniques:

1) As the number of unknown parameters increases, the amount of processing increases exponentially.

2) The Hough technique above can be used to discover any edge that can be expressed as a simple identity.

3) The generalized Hough transform can be used to discover shapes that cannot be represented by simple mathematical identities.

References

[1] Adams R and Bischof L. Seeded region growing. *IEEE Transactions on Pattern Analysis and Machine Intelligence*, 16:641-647, 1994.

[2] Affymetrix Corporation. http://www.affymetrix.com/index.affx, 2005.

[3] Agrawal R, Gehrke J, Gunopulos D, and Raghavan P. Automatic subspace clustering of high dimensional data for data mining applications. In *Proceedings of The ACM SIGMOD International Conference on Management of Data*, pages 94-105, New York: ACM Press, June 1998.

[4] Agarwala A, Dontcheva M, Agrawala M, Drucker S, Colburn A, Curless B, Salesin D, and Cohen M. Interactive digital photomontage. *ACM Transactions on Graphics*, 23(3):294-302, 2004.

[5] Akutsu T, Miyano S, and Kuhara S. Identification of genetic networks from a small number of gene expression patterns under the boolean network model, *Proc. of Pacific Symposium on Biocomputing*, 4:17-28, 1999.

[6] Anderson BDO and Moore JB. *Optimal Filtering.* Englewood Cliffs, NJ: Prentice Hall, 1979.

[7] Antoniol G and Ceccarelli M. A Markov random field approach to microarray image gridding. In: *Proceedings of the 17th International Conference on Pattern Recognition*, 3:550-553, 2004.

[8] Antoniol G, Ceccarelli M, and Petrosino A. Microarray image addressing based on the Radon transform. In *Proceedings of IEEE International Conference on Image Processing*, 1:13-16, 2005.

[9] Ahuja N, Rosenfeld A, and Haralick RM. Neighbour gray levels as features in pixel classification. *Pattern Recognition*, 12:251-260, 1980.

[10] Anderberg MR. *Cluster Analysis for Applications.* New York: Academic Press Inc, 1973.

[11] Angulo J and Serra J. Automatic analysis of DNA microarray images using mathematical morphology. *Bioinformatics*, 19(5):553-562, 2003.

[12] Appleton B and Talbot H. Globally optimal geodesic active contours. *Journal of Mathematical Imaging and Vision*, 23(1):67-86, 2005.

[13] Arabidopsis GI. Analysis of the genome sequence of the flowering plant Arabidopsis thaliana. *Nature*, 408:796-815, 2000.

[14] Axon. *GenePix© Pro 6.0 Microarray image analysis.* http://www.moleculardevices.com/pages/software/gn-genepix-pro.html, 2005.

[15] Axon. *Acuity© 4.0 Gene expression analysis.* http://www.moleculardevices.com/pages/software/gn-acuity.html, 2005.

[16] Bajcsy P. Gridline: automatic grid alignment in DNA microarray scans. *IEEE Transactions on Image Processing*, 13(1):15-25, 2004.

[17] Bajcsy P. An overview of DNA microarray grid alignment and foreground separation approaches. *EURASIP Journal on Applied Signal Processing*, 2006:1-13, Article ID 80163, 2006.

[18] Balasubramaniyan R, Hüllermeier E, Weskamp N, and Kämper J. Clustering of gene expression data using a local shape-based similarity measure. *Bioinformatics*, 21(7):1069-1077, 2005.

[19] Ball GB. Data analysis in the social sciences: what about the details? In *Procedures of Joint Computing Conference*, pages 533-559, Washington, DC: Spartan Books, 1965.

[20] Barrett WA and Cheney AS. Object-based image editing. In: *Proceedings of the 29th conference on Computer graphics and interactive techniques*, 777-784, San Antonio, July 2002.

[21] Baugh LR, Hill AA, Claggett JM, Hill-Harfe K, Wen JC, Slonim DK, Brown EL and Hunter CP. The homeodomain protein PAL-1 specifies a lineage-specific regulatory network in the *C. elegans embryo, Development*, 132:1843-1854, 2005.

[22] Beal MJ, Falciani F, Ghahramani Z, Rangel C, and Wild DL. A Bayesian approach to reconstructing genetic regulatory networks with hidden factors, *Bioinformatics*, October, 2004.

[23] Beare R and Buckley M. "Spot." http://spot.cmis.csiro.au/spot/doc/Spot.pdf.

[24] Berkhin P. *Survey of Clustering Data Mining Techniques.* Accrue Software, San Jose, CA, 2002.

[25] Bertalmio M, Bertozzi A, and Sapiro G. Navier-stokes, fluid dynamics, and image and video inpainting. In *Proceedings of the 2001 IEEE Computer Society Conference on Computer Vision and Pattern Recognition*, vol. 1, pages 355-362, Dec. 2001.

[26] Beucher S and Meyer F. The morphological approach to segmentation: the watershed transformation. In *Mathematical Morphology in Image Processing*, Chapter 12, 433-481, New York: Marcel Dekker, 1993.

[27] Bhanu B and Faugeras O. Segmentation of images having unimodal distributions. *IEEE Transactions on Pattern Analysis Machine Learning and Intelligence*, 4:408-419, 1982.

[28] BioDiscovery. AutoGene, http://www.biodiscovery.com, 2001.

[29] Biran I, Walt DR, and Epstein J. Fluorescence-based nucleic acid detection and microarrays. *Anal. Chim. Acta*, 496(1):3-36, 2002.

[30] Blekas K, Galatsanos NP, and Georgiou I. An unsupervised artifact correction approach for the analysis of DNA microarray images. In: *Proceedings of International Conference on Image Processing*, 2:165-168, 2003.

[31] Bozdech Z, Llinas M, Pulliam B, Wong ED, and Zhu J. The transcriptome of the intraerythrocytic developmental cycle of plasmodium falciparum, *PLoS Biology*, 1(1):85-100, 2003.

[32] Bozinov D and Rahenführer J. Unsupervised technique for robust target separation and analysis of DNA microarray spots through adaptive pixel clustering. *Bioinformatics*, 18(5):747-756, 2002.

[33] Bozinov D. Autonomous system for web based microarray image analysis. *IEEE Transactions on Nanobioscience*, 2(4):215-220, 2003.

[34] Brandle N, Bischof H, and Lapp H. A generic and robust approach for the analysis of spot array images. In *Proceedings of SPIE in progress in biomedical optics and imaging: microarrays: optical technologies and informatics*, 4266:1-12, San Jose, CA, January 2001.

[35] Brandle N, Chen HY, Bischof H, and Lapp H. Robust parametric and semi-parametric spot fitting for spot array images. In *Proceedings 8th Intel Systems for molecular biology*, 20-23:46-56, La Jolla, CA, August 2000.

[36] Brandle N, Lapp H, and Bischof H. Fully Automatic grid fitting for genetic spot array images containing guide spots. *ReportTechnical Report, PRIP-TR-58*, Vienna University of Technology, Austria, 2000.

[37] Buhler J, Ideker T, and Haynor D. Dapple: improved techniques for finding spots on DNA microarrays. *UV CSE Technical Report*, UWTR, 2000.

[38] Ceccarelli M and Antoniol G. A deformable grid-matching approach for microarray images. *IEEE Trans Image Process.*, 15(10):3178-3188, 2006.

[39] Chan T, Kang S, and Shen J. Euler's elastica and curvature based inpaintings. *Journal of Applied Mathematics*, 63(2):564-592, 2002.

[40] Chargaff E. Chemical specificity of nucleic acids and mechanism of their enzymic degradation. *Experientia*, 6:201-209, 1951.

[41] Chargaff E. Structure and function of nucleic acids as cell constituents. *Fed. Proceedings*, 10:654-659, 1951.

[42] Chatfield C. *The Analysis of Time Series: an Introduction*, 6th Edition, Chapman and Hall, 2004.

[43] Chen T, He HL, and Church GM. Modeling gene expression with differential equations, *Proc. of Pacific Symposium on Biocomputing*, 4:29-40, 1999.

[44] Chen Y, Dougherty ER, and Bittner ML. Ratio-based decisions and the quantitative analysis of cDNA microarray images. *Journal of Biomedical Optics*, 2(4):364-374, 1997.

[45] Cheriet M, Said JN, and Suen CY. A recursive thresholding technique for image segmentation. *IEEE Transactions on Image Processing*, 7(6):918-920, 1998.

[46] Churchill GA. Fundamentals of experimental design for cDNA microarrays. *Nature Genetics*, 32:490-495, 2002.

[47] Coe B. You want Ketchup with your DNA chips? An overview of expression microarrays. *BioTeach Online Journal*, 1(1):89-94, 2003.

[48] Cook DL, Gerber AN, and Tapscott SJ. Modeling stochastic gene expression: implications for haploinsufficiency. *Proceedings of the National Academy of Science*, USA, 95:15641-15646, 1998.

[49] Crick FH. On protein synthesis. *Symp. Soc. Exp. Biol.*, 12:138-163, 1958.

[50] Crick FH. Central dogma of molecular biology. *Nature*, 227:561-563, 1970.

[51] Crick HF, Barnett L, Brenner S, and Watts-Tobin RJ. General nature of genetic code for proteins. *Nature*, 192:1227-1232, 1961.

[52] CSIRO. SPOT Image Analysis Software. *Mathematical and Informational Sciences*, http://experimental.act.cmis.csiro.au/spot/index.php, 2005.

[53] Dave RN. Generalized fuzzy C-shells clustering and detection of circular and elliptical boundaries. *Pattern Recognition*, 25(1):713-722, 1992.

[54] Daxin J, Chun T, and Aidong Z. Cluster analysis for gene expression data: a survey. *IEEE Transactions on Knowledge and Data Engineering*, 16(11):1370-1386, 2004.

[55] de Hoon MJ, Imoto S, Kobayashi K, Ogasawara N, and Miyano S. Inferring gene regulatory networks from time-ordered gene expression data of bacillus subtilis using differential equations, *Proc. of Pacific Symposium on Biocomputing*, 17-28, 2003.

[56] de Jong H. Modeling and simulation of genetic regulatory systems: a literature review, *J Comput Biol*, 9(1):67-103, 2002.

[57] Deng N and Duan H. An automatic and power spectra-based rotate correcting algorithm for microarray image. *Proceedings of IEEE Eng Med Biol Soc Conference*, 1:898-901, 2005.

[58] D'haeseleer P, Wen X, Fuhrman S, and Somogyi R. Linear modeling of mRNA expression levels during CNS development and injury, *Proc. of Pacific Symposium on Biocomputing*, 41-52, 1999.

[59] Diday E and Simon JC. Clustering analysis. *Digital Pattern Recognition*, 1(1):47-94, Berlin: Springer-Verlag, 1976.

[60] Diggle P. *Time Series: a Biostatistical Introduction*, Oxford Statistical Science Series 5, 1990.

[61] DeRisi JL, Iyer VR, and Brown PO. Exploring the metabolic and genetic control of gene expression on a genomic scale. *Science*, 278(5338):680-686, 1996.

[62] DeRisi JL, Penland L, Brown PO, Bittner ML, Meltzer PS, Ray M, Chen YD, Su YA, and Trent JM. Use of a cDNA microarray to analyse gene expression patterns in human cancer. *Nature Genetics*, 14(4):457-460, 1996.

[63] Dubes RC. *Cluster Analysis and Related Issues*, 1st ed. River Edge, NJ: World Scientific Publishing Co. Inc., 1993.

[64] Dumochel W. Data squashing: constructing summary datasets. In Abell OJ, Pardalos P, and Resende GC, editors, *Handbook of Massive Datasets*, 1st ed. 579-592, Amsterdam, Kluwer Academic Publishers, 2002.

[65] Dumochel W, Volinsky C, Johnson T, Cortes C, and Pregibon D. Squashing flat files flatter. In *Proceedings of the 5th ACM Conference on Knowledge Discovery and Data Mining*, 6-15, New York: ACM Press, August 1999.

[66] Dunn CJ. A fuzzy relative of ISODATA process and its use in detecting compact well-separated clusters. *Journal of Cybernetics*, 3(3):32-57, 1974.

[67] Duyk GM. Sharper tools and simpler methods. *Nature Genetics*, 32:474-479, 2002.

[68] Efros AA and Leung TK. Texture synthesis by non-parametric sampling. In *Proceedings of the IEEE International Conference on Computer Vision*, 1033-1038, 1999.

[69] Eisen M. *ScanAlyse*. http://rana.lbl.gov/EisenSoftware.htm, 2002.

[70] Eisen M, Spellman PT, Brown PO, and Botstein D. Cluster analysis and display of genome-wide expression Ppatterns. In *Proceedings of the National Academy of Sciences*, 14863-14868, Washington, DC, December 1998.

[71] Ekins RP. *UK Patent Application 8803000*, 1987.

[72] Ekins RP, Chu R, and Biggart E. The development of microspot, multi-analyte radiometric immunoassay using dual fluorescent-labelled antibodies. *Anal. Chim. Acta.*, 227:73-96, 1989.

[73] Ekins RP and Chu FW. Multianalyte microspot immunoassay-microanalytical "compact disk" of the future. *Clinical Chemistry*, 37:1955-1967, 1991.

[74] Ekins R, Chu F, and Micallef. J. High specific activity chemiluminescent and fluorescent markers: their potential application to high sensitivity and "multianalyte" immunoassays. *J. Journal Biolumin. Chemilumin*, 4:59-78, 1989.

[75] Epstein J, Leung A, Lee K, and Walt DR. High-density, microsphere-based fiber optic DNA microarrays. *Biosensors and Bioelectronics*, 5-6(18):541-546, 2003.

[76] Fabbri R, da Costa LF, and Barrera J. Towards non-parametric gridding of microarray images. In *14th International Conference on Digital Signal Processing*, 2:623-626, 2002.

[77] Fleischmann RD, Adams MD, White O, Clayton RA, Kirkness EF, Kerlavage AR, and Bult CJ. Whole genome random sequencing and assembly of Haemophilus influenzae Rd. *Science*, 269:496-512, 1995.

[78] Fodor SP, Read JL, Pirrung MC, Stryer L, Lu AT, and Solas D. Light-directed, spatially addressable parallel chemical synthesis. *Science*, 251(4995):767-773, 1991.

[79] Fodor SP, Rava RP, Huang XC, Pease AC, Holmes CP, and Asams CL. Multiplexed biochemical assays with biological chips. *Nature*, 364:555-556, 1993.

[80] Forgy E. Cluster analysis of multivariate data: efficency vs. interpretability of classification. *Biometrics*, 21:768, 1965.

[81] Franklin RE and Gosling RG. Evidence for a 2-chain helix in the crystalline structure of sodium deoxyribonucleate. *Nature*, 172:156-157, 1953.

[82] Franklin RE and Gosling RG. The structure of sodium thymonucleate fibres: I. The influence of water content. II. The cylindrically symmetrical Patterson function. *Acta Crystallographica*, 6:673-685, 1953.

[83] Fraser CM, Gocayne JD, White O, Adams MD, Clayton RA, Fleischmann RD, Bult CJ, Kerlavage AR, Sutton G, Kelley JM, Fritchman RD, Weidman JF, Small KV, Sandusky M, Fuhrmann J, Nguyen D, Utterback TR, Saudek DM, Phillips CA, Merrick JM, Tomb JF, Dougherty BA, Bott KF, Hu PC, Lucier TS, Peterson SN, Smith HO, Hutchison CA III, and Venter JC. The minimal gene complement of Mycoplasma genitalium. *Science*, 270:397-403, 1995.

[84] Fraser K, O'Neill P, Wang Z, and Liu X. Copasetic analysis: a framework for the blind analysis of microarray imagery. *Systems Biology*, 1(1):190-196, 2004.

[85] Fraser K, O'Neill P, Wang Z, and Liu X. Copasetic analysis: automated analysis of biological gene expression images. *International Conference on Control, Automation, Robotics and Vision*, 1061-1066, December 6-9, 2004.

[86] Fraser K, O'Neill P, Wang Z, and Liu X. Copasetic clustering of cDNA microarray imagery. In Mahammadian M. ed., *International Conference on Computational Intelligence for Modelling Control and Automation*, pages 779-787, July 12-14, 2004.

[87] Fraser K, O'Neill P, Wang Z, and Liu X. Copasetic clustering: making sense of large-scale images. *Lecture Notes in Computer Science*, 3327:99-108, 2004.

[88] Fraser K, Wang Z, Li Y, Kellam P, and Liu X. Improving microarray expressions with recalibration. In: *Proc. International Symposium on Computational Life Science*, 3-16, the Netherlands, October 2007.

[89] Fraser K, Wang Z, Li Y, Kellam P, and Liu X. Noise filtering and microarray image reconstruction via chained fouriers. *Lecture Notes in Computer Science*, 4723:308-319, 2007.

[90] Friedman N, Linial M, Nachman I, and Pe'er D. Using Bayesian networks to analyse expression data. In *Proceedings ACM Research in Computational Molecular Biology (RECOMB)*, 127-135, Tokyo, Japan, 2000.

[91] Frigui H and Krishnapuram RA. Robust competitive clustering algorithm with applications in computer vision. *IEEE Transactions on Pattern Analysis and Machine Intelligence*, 21(5):450-465, 1999.

[92] Gasch AP, Spellman PT, Kao CM, Carmel-Harel O, Eisen MB, Storz G, Botstein D, and Bown PO. Genomic expression program in the response of yeast cells to environmental changes. *Molecular Biology of the Cell*, 11:4241-4257, 2000.

[93] Geller CS, Gregg PJ, Hagerman P, and Rocke MD. Transformation and normalization of oligonucleotide microarray data. *Bioinformatics*, 19(14):1817-1823, 2003.

[94] Gers F, Schraudolph N and Schmidhuber J. Learning precise timing with LSTM recurrent networks, *Journal of Machine Learning Research*, 3:115-143, 2002.

[95] Ghahramani Z and Hinton GE. Parameter estimation for linear dynamical systems, *Technical Report*, University of Toronto, 1996.

[96] Ghahramani Z. Learning Dynamic Bayesian Networks. In C.L. Giles and M. Gori (eds.), *Adaptive Processing of Sequences and Data Structures*. Lecture Notes in Artificial Intelligence, 168-197. Berlin: Springer-Verlag.

[97] Giannakeas N, Fotiadis DI, and Politou AS. An automated method for gridding in microarray images. In *Proceedings of 28th Annual International Conference of the IEEE Engineering in Medicine and Biology Society*, 5876-5879, 2006.

[98] Gonzalez RC and Woods RE. *Digital Image Processing*. New York: Prentice Hall, 2nd ed, 2002.

[99] Hand DJ, Mannila H, and Smyth P. *Principles of Data Mining*, Cambridge, MA: MIT Press, 2001.

[100] Hartelius K and Carstensen JM. Bayesian grid matching. *IEEE Transactions on Pattern Analysis and Machine Intelligence*, 25(2):162-173, 2003.

[101] Hershey AD and Chase M. Independent functions of viral protein and nucleic acid in growth of bacteriophage. *Journal General Physiology*, 36:39-56, 1952.

[102] Hillier LW, Miller W, Birney E, Warren W, Hardison RC, Ponting CP, Bork P, Burt DW, Groenen MA, Delany ME, Dodgson JB, Chinwalla AT, Cliften PF, Clifton SW, Delehaunty KD, Fronick C, Fulton RS, Graves TA, Kremitzki C, Layman D, Magrini V, McPherson JD, Miner TL, Minx P, Nash WE, Nhan MN, Nelson JO, Oddy LG, Pohl CS, Randall-Maher J, Smith SM, Wallis JW, Yang SP, Romanov MN, Rondelli CM, Paton B, Smith J, Morrice D, Daniels L, Tempest HG, Robertson L, Masabanda, JS, Griffin DK, Vignal A, Fillon V, Jacobbson L, Kerje S, Andersson L, Crooijmans RP, Aerts J, van der Poel JJ, Ellegren H, Caldwell RB, Hubbard SJ, Grafham DV, Kierzek AM, McLaren SR, Overton IM, Arakawa H, Beattie KJ, Bezzubov Y, Boardman PE, Bonfield JK, Croning MD, Davies RM, Francis MD, Humphray SJ, Scott CE, Taylor RG, Tickle C, Brown WR, Rogers J, Buerstedde JM, Wilson SA, Stubbs L, Ovcharenko I, Gordon L, Lucas S, Miller MM, Inoko H, Shiina T, Kaufman J, Salomonsen J, Skjoedt K, Wong GK, Wang J, Liu B, Wang J, Yu J, Yang H, Nefedov M, Koriabine M, Dejong PJ, Goodstadt L, Webber C, Dickens NJ, Letunic I, Suyama M, Torrents D, von Mering C, Zdobnov EM, Makova K, Nekrutenko A, Elnitski

L, Eswara P, King DC, Yang S, Tyekucheva S, Radakrishnan A, Harris RS, Chiaromonte F, Taylor J, He J, Rijnkels M, Griffiths-Jones S, Ureta-Vidal A, Hoffman MM, Severin J, Searle SM, Law AS, Speed D, Waddington D, Cheng Z, Tuzun E, Eichler E, Bao Z, Flicek P, Shteynberg DD, Brent MR, Bye JM, Huckle EJ, Chatterji S, Dewey C, Pachter L, Kouranov A, Mourelatos Z, Hatzigeorgiou AG, Paterson AH, Ivarie R, Brandstrom M, Axelsson E, Backstrom N, Berlin S, Webster MT, Pourquie O, Reymond A, Ucla C, Antonarakis SE, Long M, Emerson JJ, Betran E, Dupanloup I, Kaessmann H, Hinrichs AS, Bejerano G, Furey TS, Harte RA, Raney B, Siepel A, Kent WJ, Haussler D, Eyras E, Castelo R, Abril JF, Castellano S, Camara F, Parra G, Guigo R, Bourque G, Tesler G, Pevzner PA, Smit A, Fulton LA, Mardis ER, and Wilson RK. International Chicken Genome Sequencing Consortium. 2004. Sequence and comparative analysis of the chicken genome provide unique perspectives on vertebrate evolution. *Nature*, 432:695-716, 2004.

[103] Hirata R Jr, Barrera J, Hashimoto RF, and Dantas DO. Microarray gridding by mathematical morphology. In *Proceedings of 14th Brazilian Symposium on Computer Graphics and Image Processing*, 112-119, 2001.

[104] Holloway AJ, Van Laar, Ryan K, Tothill RW, and Bowtell DL. Options available from start to finish for obtaining data from DNA microarrays II. *Nature Genetics*, 32:481-489, 2002.

[105] Holter NS, Maritan A, Cieplak M, Fedoroff NV, and Banavar JR. Dynamic modeling of gene expression data, *Proceedings of the National Academy of Science*, USA, 98:1693-1698, 2001.

[106] Huang S. Gene expression profiling, genetic networks, and cellular states: an integrating concept for tumorigenesis and drug discovery,*Journal of Molecular Medicine*, 77:469-480, 1999.

[107] Hughes T, Marton M, Jones A, Roberts C, Stoughton R, Armour C, Bennett H, Coffey E, Dai H, and He Y. Functional discovery via a compendium of expression profiles. *Cell*, 102(1):109-126, 2000.

[108] International Human Genome Sequencing Consortium. 2004. Finishing the euchromatic sequence of the human genome. *Nature*, 431:931-945, 2004.

[109] Jain AK and Dubes RC. *Algorithms For Clustering Data*, 1st ed. Englewood Cliffs, NJ: Prentice Hall, 1988.

[110] Jain AK, Murty MN, and Flynn PJ. Data clustering: a review. *ACM Computing Surveys*, 31(3):264-323, 1999.

[111] Jain NA, Tokuyasu AT, Snijders MA, Segraves R, Albertson GD, and Pinkel D. Fully automated quantification of microarray image data. *Genome Research*, 12(2):325-332, 2002.

[112] Jancey RC. Multidimensional group analysis. *Australian Journal of Botany*, 14:127-130, 1966.

[113] Jansen R, Greenbaum D, and Gerstein M. Relating whole genome expression data with protein-protein interactions. *Genome Research*, 12:37-46, 2002.

[114] Jung HY and Cho HG. An automatic block and spot indexing with k-nearest neighbors graph for microarray image analysis. *Bioinformatics*, 18(Suppl 2):S141-S151, 2002.

[115] Katzer M, Kummbert F, and Sagerer G. A Markov random field model of microarray gridding. In *Proceedings of the 18th ACM Symposium on Applied Computing*, 72-77, New York: ACM Press, March, 2003.

[116] Katzer M, Kummbert F, and Sagerer G. Methods for automatic microarray image segmentation. *IEEE Transactions on Nanobioscience*, 2(4):202-214, 2003.

[117] Katzer M, Kummert F, and Sagerer G. Robust automatic microarray image analysis. In *Proceedings International Conference on Bioinformatics: North-South Networking*, Bangkok, Thailand, 2002.

[118] Kaufmann L and Rousseeuw JP. Finding groups in data: an introduction to cluster analysis. *Clustering Large Applications (Program CLARA)*, pages 126-163, New York: John Wiley & Sons, 1990.

[119] Kellam P, Liu X, Martin N, Orengo CA, Swift S, and Tucker A. A framework for modelling virus gene expression data. *Journal of Intelligent Data Analysis*, 6(3):265-280, 2002.

[120] Kepler BM, Crosby L, and Morgan TK. Normalization and analysis of DNA microarray data by self-consistency and local regression. *Genome Biology*, 3(7):research0037.1-research0037.12, 2002.

[121] Kepler TB and Elston TC. Stochasticity in transcriptional regulation: origins, consequences, and mathematical representations,*Biophys J*, 81(6):3116-3136, 2001.

[122] King B. Step-wise clustering procedures. *Journal of the American Statistical Association*, 69:86-101, 1967.

[123] Kooperberg C, Fazzio TG, Delrow JJ, and Tsukiyama T. Improved background correction for spotted cDNA microarrays. *Journal of Computational Biology*, 9:55-66, 2002.

[124] Kwatra V, Schödl A, Essa I, Turk G, and Bobick A. Graphcut Textures: image and video synthesis using graph cuts. In *Proceedings of the 2003 ACM SIGGRAPH Conference*, 277-286, San Diego, CA, July 2003.

[125] Larsen B and Aone C. Fast and effective text mining using linear-time document clustering. In Chaudhuri S, Madigan D, and Fayyad U, editors, In *Proceedings of the Fifth ACM SIGKDD International Conference on Knowledge Discovery*, 16-22, New York: ACM Press, August, 15-18, 1999.

[126] Lashkari DA, DeRisi JL, McCusker JH, Namath AF, Gentile C, Hwang SY, Brown PO, and Davis RW. Yeast microarrays for genome wide parallel genetic and gene analysis. In *Proceedings National Academy of Sciences*, 95:13057-13062, Washington, DC, 1997.

[127] Lawrence ND, Milo M, Niranjan M, Rashbass P, and Soullier S. Reducing the variability in cDNA microarray image processing by Bayesian inference. *Bioinformatics*, 20(4):518-526, 2004.

[128] Leach S and Hunter L. Comparative study of clustering techniques for gene expression microarray data. In Miyano S, Shamir R, and Akagi T, editors, *Currents in Computational Molecular Biology*, 1-2, Tokyo: University Academy Press, 2000.

[129] Leavers VF. *Shape Detection in Computer Vision Using the Hough Transform*, Berlin: Springer, 1992.

[130] Li QH, Fraley C, Bumgarner RE, Yeung KY, and Raftery AE. Donuts, scratches and blanks: Robust model-based segmentation of microarray images. *Bioinformatics*, 21(12):2875-2882, 2005.

[131] Liang S, Fuhrman S and Somogyi R. REVEAL: A general reverse engineering algorithm for inference of genetic network architectures, *Proc. of Pacific Symposium on Biocomputing*, 3:18-29, 1998.

[132] Liu T, Sung W and Mittal A. Model gene network by semi-fixed Bayesian network, *Expert Systems with Applications*, 30(1):42-49, 2006.

[133] Liu X and Kellam P. Mining gene expression data. In Orengo CA, Jones DT, and Thornton JM, eds., *Bioinformatics: Genes, Proteins and Computers*, 1st ed. 229-244, Oxford: BIOS Scientific Publishers, 2003.

[134] Ljung L. *System Identification - Theory for the User*, 2nd ed, PTR Upper Saddle River, N.J: Prentice Hall, 1999.

[135] Lonardi S and Luo Y. Gridding and compression of microarray images. *Proc IEEE Comput Syst Bioinform Conf.*, 122-130, 2004.

[136] McAdams HM and Arkin A. Stochastic mechanisms in gene expression. *Proceedings of the National Academy of Science*, Washington, DC, 94:814-819, 1997.

[137] Maduro MF and Rothman JH. Making worm guts: the gene regulatory network of the Caenorhabditis elegans endoderm, *Dev. Biol.*, 246:68-85, 2002.

[138] Matsuzaki M, Misumi O, Shin-I T, Maruyama S, Takahara M, Miyag-ishima SY, Mori T, Nishida K, Yagisawa F, Yoshida Y, Nishimura Y, Nakao S, Kobayashi T, Momoyama Y, Higashiyama T, Minoda A, Sano M, Nomoto H, Oishi K, Hayashi H, Ohta F, Nishizaka S, Haga S, Miura S, Morishita T, Kabeya Y, Terasawa K, Suzuki Y, Ishii Y, Asakawa S, Takano H, Ohta N, Kuroiwa H, Tanaka K, Shimizu N, Sugano S, Sato N, Nozaki H, Ogasawara N, Kohara Y, and Kuroiwa T. Genome se-quence of the ultrasmall unicellular red alga Cyanidioschyzon merolae 10D. *Nature*, 428:653-657, 2004.

[139] McQueen J. Some methods for classification and analysis of multivariate observations. In Le Cams L and Neyman S, eds., *Proceedings of the 5th Berkeley Symposium on Mathematics Statistics and Probability*, 281-297, 1967.

[140] Mehnert A and Jackway P. An improved seeded region growing algo-rithm. *Pattern Recognition Letters*, 18:1065-1071, 1997.

[141] Mitra S and Hayashi Y. Bioinformatics with soft computing. *IEEE Transactions on Systems, Man and Cybernetics - Part C*, 36(5):616-635, 2006.

[142] Moore SK. Making chips. *IEEE Spectrum*, 38(3):54-60, 2001.

[143] Motwani R and Raghavan P. *Randomized Algorithms*, Cambridge, UK and New York: Cambridge University Press, 1995.

[144] Mouse Genome Sequencing Consortium. Initial sequencing and compar-ative analysis of the mouse genome. *Nature*, 420:520-562, 2002.

[145] Mullis KB and Faloona FA. Specific synthesis of DNA in vitro via a polymerase-catalyzed chain reaction. *Methods Enzymol*, 155:335-350, 1987.

[146] Murphy K and Mian S. Modelling gene expression data using dynamic Bayesian networks. *Technical Report*, Berkeley: University of California, 1999.

[147] Murtagh F. A survey of recent advances in hierarchical clustering algo-rithms which use cluster centers. *The Computer Journal*, 26(4):354-359, 1983.

[148] Murty MN and Krishna G. A computationally efficient technique for data clustering. *Pattern Recognition*, 12:153-158, 1980.

[149] Nagarajan R and Peterson CA. Identifying genes in microarray images. *IEEE Transactions on Nanobioscience*, 1(2):78-84, 2002.

[150] Netravali AN and Haskell BG. *Digital pictures: representation, compres-sion and standards*, 2nd ed., New York: Plenum Press, 1995.

[151] Ng R and Han J. Efficient and effective clustering method for spatial data mining. In *Proceedings of the 20th International Conference on Very Large Data Bases*, 144-155, September 1994.

[152] Niblack W. *An Introduction to Digital Image Processing*, London: Prentice Hall international Ltd., 1986.

[153] Nielsen F and Nock R, ClickRemoval: interactive pinpoint image object removal. In *Proceedings of the 13th annual ACM international conference on Multimedia*, 315-318, Hilton, Singapore, Nov. 2005.

[154] Nirenberg MW and Matthaei HJ. The dependence of cell-free protein synthesis in E. coli upon naturally occurring or synthetic polyribonucleotides. In *Proceedings National Academy of Sciences*, 47:1589, Washington, DC, 1961.

[155] Oliveira MM, Bowen B, McKenna R, and Chang YS. Fast digital image inpainting. In: *Proceedings of the Visualization, Imaging and Image Processing*, Marbella, Spain, September 2001, 261-266.

[156] O'Neill P, Fraser K, Wang Z, Kellam P, Kok JN, and Liu X. Pyramidic clustering of large scale microarray imagery. *The Computer Journal*, 48:466-479, 2005.

[157] O'Neill P, Magoulas GD, and Liu X. Improved processing of microarray data using image reconstruction techniques. *IEEE Transactions on Nanobioscience*, 2(4):176-183, 2003.

[158] Otsu N. A threshold selection method from grey level histograms. *IEEE Transactions on Systems, Man and Cybernetics*, 8:62-66, 1978.

[159] Pal SK, Bandyopadhyay S, and Ray SS, Evolutionary computation in bioinformatics: a review. *IEEE Transactions on Systems, Man and Cybernetics - Part C*, 36(5):601-615, 2006.

[160] Pauling L and Corey RB. Two Hydrogen-bonded spiral configurations of the polypeptide chain. *Journal American Chemical Society*, 72:5349, 1950.

[161] Pauling L and Corey RB. Stable configurations of polypeptide chains. In *Proceedings Royal Society*, 141:21-33, London, 1953.

[162] Pauling L, Corey RB, and Branson HR. The structure of proteins, two hydrogen-bonded helical configurations of the polypeptide chain. In *Proceedings National Academy Sciences*, 37:205-511, Washington, DC, 1951.

[163] Pease AC, Solas D, Sullivan J, Cronin MT, Holmes CP, and Fodor SP, Light-generated oligonucleotide arrays for rapid DNA sequence analysis. In *Proceedings National Academy Sciences*, 91(11):5022-5026, Washington, DC, 1994.

[164] Perez P, Gangnet M, and Blake A. Poisson image editing. *ACM Transactions on Graphics*, 22(3):313-318, 2003.

[165] Perkins WA. Area segmentation of images using edge points. *IEEE Transactions on Pattern Recognition and Machine Intelligence*, 2(1):8-15, 1980.

[166] Petricoin EF, III, Emanuel F, Hackett JL, Lesko LJ, Puri RK, Gutman SI, Chumakov K, Woodcock J, Feigal JR, David W, Zoon CK, and Sistare DF. Medical applications of microarray technologies: a regulatory science perspective. *Nature Genetics*, 32:474-479, 2002.

[167] Quackenbush J. Computational analysis of microarray analysis. *Nature Reviews Genetics*, 2(6):418-427, 2001.

[168] Quackenbush J. Microarray data normalization and transformation. *Nature Genetics*, 32:490-495, 2002.

[169] QuantArray. GSI Lumonics. http://www.bipl.ahc.umn.edu/quantarray.html, 2005.

[170] Rahnenführer J and Bozinov D. Hybrid clustering for microarray image analysis combining intensity and shape features. *BMC Bioinformatics*, 5(47), 2004.

[171] Ramoni MF, Sebastiani P, and Kohane IS. Cluster analysis of gene expression dynamics, *Proceedings of the National Academy of Science*, USA, 99:9121-9126, 2002.

[172] Ranada S and Rosenfeld A. Point pattern matching by relaxation. *Pattern Recognition*, 12:269-275, 1980.

[173] Rangel C, Angus J, Ghahramani Z, Lioumi M, Sotheran EA, Gaiba A, Wild DL, and Falciani F. Modeling T-cell activation using gene expression profiling and state space models, *Bioinformatics*, 20(9):1361-1372, 2004.

[174] Rother C, Kolmogorov V, and Blake A. GrabCut - Interactive foreground extraction using iterated graph cuts. In *Proceedings of the 2004 ACM SIGGRAPH conference*, 309-314, Los Angeles, CA, August 2004.

[175] Rueda L and Vidyadharan V. A hill-climbing approach for automatic gridding of cDNA microarray images. *IEEE/ACM Trans Comput Biol Bioinform*, 3(1):72-83.

[176] Ruspini EH. A new approach to clustering. *Information Control*, 15(1):22-32, 1969.

[177] Russell S and Norvig P. *Artificial Intelligence A Modern Approach*, Upper Saddle River, NJ: Prentice Hall, 2nd ed., 2003.

[178] Samartzidou H, Turner L, Houts T, Frome J, Worley M, and Albertsen H. Lucidea microarray scorecard: an integrated analysis tool for microarray experiments, *Life Science News*, 7(13):1-10, 2001.

[179] Sanger F, Nicklen S, and Coulson AR. DNA sequencing with chain-terminating inhibitors. In *Proceedings National Academy Sciences*, 74:5463-5467, 1977.

[180] Sanger F, Air GM, Barrell BG, Brown NL, Coulson AR, Fiddes CA, Hutchison CA, Slocombe PM, and Smith M. Nucleotide sequence of bacteriophage phi X174 DNA. *Nature*, 165:687-695, 1977.

[181] Sanger F, Coulson AR, Friedmann T, Air GM, Barrell BG, Brown NL, Fiddes JC, Hutchison CA 3rd, Slocombe PM, and Smith M. The nucleotide sequence of bacteriophage phi-X174. *Journal Molecular. Biology*, 125:225-246, 1977.

[182] Sanger F, Coulson AR, Barrell BG, Smith AJ, and Roe BA. Cloning in single stranded bacteriophage as an aid to rapid DNA sequencing. *Journal Molecular Biology*, 143:161-178, 1980.

[183] Scanalytics, Inc. MicroArray Suite, http://www.scanalytics.com/products/hts/microarray.html, 2005.

[184] ScanAlyze Manual, http://rana.lbl.gov/manuals/ScanAlyzeDoc.pdf, Online Resource, 2007.

[185] Schena M and Davis RW. In *PCR Methods Manual*, New York: Academic Press (in press).

[186] Schena M, Shalon D, Davis RW, and Brown OP. Quantitative monitoring of gene expression patterns with a complementary DNA microarray. *Science*, 210:467-470, 1995.

[187] Schena D, Shalon R, Heller AC, Brown PO, and Davis RW. Parallel human genome analysis: microarray based expression monitoring of 1000 genes. In *Proceedings National Academy Sciences*, 93(20):10614-10619, USA, 1996.

[188] Shumway RH and Stoffer DS. An approach to time series smoothing and forecasting using the EM algorithm, *J. Time Series Anal.*, 3:253-264, 1982.

[189] Smith LM. Fluorescence detection in automated DNA sequence analysis. *Nature*, 321:674-679, 1986.

[190] Smolen P, Baxter DA and Byrne JH. Mathematical modeling of gene networks review, *Neuron*, 26(3):567-580, 2000.

[191] Smyth KG, Yang HY, and Speed PT. Statistical issues in cDNA microarray data analysis. *Functional Genomics: Methods and Protocols*, 224:111-136, 2003.

[192] Sneath PH and Sokal RR. *Numerical Taxonomy – The Principles and Practice of Numerical Classification*, San Francisco: W. H. Freeman, 1973.

[193] Soille P. *Morphological Image Analysis: Principles and Applications*, New York: Springer, 1999.

[194] Somogyi R and Sniegoski CA. Modeling the complexity of genetic networks: Understanding multigenic and pleiotropic regulation, *Complexity*, 1(6):45-63, 1996.

[195] Southern EM. Detection of specific sequences among DNA fragments separated by gel electrophoresis. *Journal Molecular Biology*, 98:503-517, 1975.

[196] Spellman PT, Sherlock G, Zhang MQ, Iyer VR, Anders K, Eisen MB, Brown PO, Botstein D, and Futcher B. Comprehensive identification of cell cycle-regulated genes of the yeast Saccharomyces cerevisiae by microarray hybridization. *Molecular biology of the cell*, 9(12):3273-3297, 1998.

[197] Starck J, Murtagh FD, and Bijaoui A. *Image Processing and Data Analysis: The Multiscale Approach*, Cambridge, UK and New York: Cambridge University Press, 1998.

[198] Steinbach M, Karypis G, and Kumar V. A comparison of document clustering techniques. In *Proceedings of the 6th ACM SIGKDD World Text Mining Conference*, 1-20, New York: ACM Press, August 20-23, 2000.

[199] Steinbath M, Wruck W, Seidel H, Lehrach H, Radelof U, and O'Brien J. Automated image analysis for hybridization experiments. *Bioinformatics*, 17:634-641, 2001.

[200] Sun J, Yuan L, Jia J, and Shum H. Image completion with structure propagation. In *Proceedings of the 2005 ACM SIGGRAPH Conference*, 861-868, New York: ACM Press, 2005.

[201] Swift S and Liu X. Predicting glaucomatous visual field deterioration through short multivariate time series modelling, *Artificial Intelligence in Medicine*, 24:5-24, 2002.

[202] Tamayo P, Slonim D, Mesirov J, Zhu Q, Kitareewan S, Dmitrovsky E, Lander ES and Golub TR. Interpreting patterns of gene expression with self-organizing maps: Methods and application to hematopoietic differentiation, *Proceedings of the National Academy of Science*, Washington, DC, 96:2907-2912, 1999.

[203] Tavazoie S, Hughes JD, Campbell MJ, Cho RJ and Church GM. Systematic determination of genetic network architecture, *Nature Genetics*, 22(3):281-285, 1999.

[204] Thattai T and van Oudenaarden A. Stochastic Gene Expression in Fluctuating Environments, *Proc. of the Genetics Society of America*, 523-530, 2004.

[205] The Chipping Forcast I. *Nature Genetics*, 21(Supplement):1-60, 1999.

[206] The Chipping Forcast II. *Nature Genetics*, 32:461-552, 2002.

[207] Tian T and Burrage K. Stochastic neural network models for gene regulatory networks, *Proc. of the 2003 IEEE Congress on Evolutionary Computation*, 162-169, 2003.

[208] Vincent L and Soille P. Watersheds in digital spaces: an efficient algorithm based on immersion simulations. *IEEE Transactions on Pattern Analysis and Machine Intelligence*, 13:583-598, 1991.

[209] Wang XH, Istepanian RSH, and Song YH. Application of wavelet modulus maxima in microarray spots recognition. *IEEE Transactions on Nanobioscience*, 2(4):190-192, 2003.

[210] Wang XH, Istepanian RSH, and Song YH. Microarray image enhancement by denoising using stationary wavelet transform. *IEEE Transactions on Nanobioscience*, 2(4):184-189, 2003.

[211] Wang YK and Huang CW. DNA microarray image analysis using active contour model. In *Proceedings of IEEE Computational Systems Bioinformatics Conference*, 549-550, 2004.

[212] Wann CD and Thomopoulos SA. A Comparative study of self-organizing clustering algorithms Dignet and ART2. *Neural Networks*, 10(4):737-743, 1997.

[213] Ward JH. Hierarchical grouping to optimize an objective function. *Journal of American Statistical Association*, 58:236-244, 1963.

[214] Watson JD. Involvement of RNA in the synthesis of proteins. *Science*, 140:17-26, 1963.

[215] Watson JD and Crick FH. Genetic implications of the structure of deoxyribonucleic acid. *Nature*, 171:964-967, 1953.

[216] Watson JD and Crick FH. Molecular structure of nucleic acids: a structure for deoxyribonucleic acid. *Nature*, 171:737-738, 1953.

[217] Wang Z, Lam J, Wei G, Fraser K, and Liu X. Filtering for nonlinear genetic regulatory networks with stochastic disturbances. *IEEE Transactions on Automatic Control*, 53(10):2448-2457, 2008.

[218] Wang Z, Gao H, Cao J, and Liu X. On delayed genetic regulatory networks with polytopic uncertainties: robust stability analysis. *IEEE Transactions on NanoBioscience*, 7(2):154-163, 2008.

[219] Wang Z, Yang F, Ho DWC, Swift S, Tucker A, and Liu X. Stochastic dynamic modeling of short gene expression time series data. *IEEE Transactions on NanoBioscience*, 7(1):44-55, 2008.

[220] Weinstein E, Oppenheim AV, Feder M, and Buck JR. Iterative and sequential algorithms for multisensor signal enhancement, *IEEE Trans. Signal Processing*, 42:846-859, 1994.

[221] Wessels LF, Someren EPvan, and Reinders MJ. A comparison of genetic network models, *Proc. of Pacific Symposium on Biocomputing*, 508-519, 2001.

[222] Wu F, Zhang W, and Kusalik AJ. Modeling gene expression from microarray expression data with state-space equations, *Proc. of Pacific Symposium on Biocomputing*, 581-592. Hawaii, USA, 2004.

[223] Xia Q, Zhou Z, Lu C, Cheng D, Dai F, Li B, Zhao P, Zha X, Cheng T, Chai C, Pan G, Xu J, Liu C, Lin Y, Qian J, Hou Y, Wu Z, Li G, Pan M, Li C, Shen Y, Lan X, Yuan L, Li T, Xu H, Yang G, Wan Y, Zhu Y, Yu M, Shen W, Wu D, Xiang Z, Yu J, Wang J, Li R, Shi J, Li H, Li G, Su J, Wang X, Li G, Zhang Z, Wu Q, Li J, Zhang Q, Wei N, Xu J, Sun H, Dong L, Liu D, Zhao S, Zhao X, Meng Q, Lan F, Huang X, Li Y, Fang L, Li C, Li D, Sun Y, Zhang Z, Yang Z, Huang Y, Xi Y, Qi Q, He D, Huang H, Zhang X, Wang Z, Li W, Cao Y, Yu Y, Yu H, Li J, Ye J, Chen H, Zhou Y, Liu B, Wang J, Ye J, Ji H, Li S, Ni P, Zhang J, Zhang Y, Zheng H, Mao B, Wang W, Ye C, Li S, Wang J, Wong GK, and Yang H. A draft sequence for the genome of the domesticated silkworm (Bombyx mori). *Science*, 306:1937-1940, 2004.

[224] Xu C and Prince JL. Snakes, shapes, and gradient vector flow. *IEEE Transactions on Image Processing*, 7(3):359-369, 1998.

[225] Xu R and Wunsch D. Survey of clustering algorithms. *IEEE Transactions on Neural Networks*, 16(3):645-678, 2005.

[226] Yang HY, Dudoit S, Luu P, and Speed PT. Normalisation for cDNA Microarray Data, 2000.

[227] Yang YH, Buckley MJ, Dudoit S, and Speed TP. Comparison of methods for image analysis on cDNA microarray data. *Journal of Computational and Graphical Statistics*, 11:108-136, 2002.

[228] Yeung KY, Fraley C, Murua A, Raftery AE, and Ruzzo WL. Model-based clustering and data transformations for gene expression data, *Bioinformatics*, 17(10):977-987, 2001.

[229] Zadeh LA. Fuzzy sets. *Information Control*, 8(1):338-353, 1965.

[230] Zhang T, Ramakrishnan R, and Livny M. BIRCH: An efficient data clustering method for very large databases. In *Proceedings of the ACM*

SIGMOD International Conference on Management of Data, 103-114, New York: ACM Press, June, 1996.

[231] Ziskind I and Hertz D. Maximum-likelihood localization of narrow-band autoregressive sources via the EM algorithm, *IEEE Trans. Signal Processing*, 41(8):2719-2724, 1993.

Index